T0226373

SpringerBriefs in Applied Sciences and Technology

SpringerBriefs present concise summaries of cutting-edge research and practical applications across a wide spectrum of fields. Featuring compact volumes of 50–125 pages, the series covers a range of content from professional to academic.

Typical publications can be:

- A timely report of state-of-the art methods
- An introduction to or a manual for the application of mathematical or computer techniques
- A bridge between new research results, as published in journal articles
- A snapshot of a hot or emerging topic
- An in-depth case study
- A presentation of core concepts that students must understand in order to make independent contributions

SpringerBriefs are characterized by fast, global electronic dissemination, standard publishing contracts, standardized manuscript preparation and formatting guidelines, and expedited production schedules.

On the one hand, **SpringerBriefs in Applied Sciences and Technology** are devoted to the publication of fundamentals and applications within the different classical engineering disciplines as well as in interdisciplinary fields that recently emerged between these areas. On the other hand, as the boundary separating fundamental research and applied technology is more and more dissolving, this series is particularly open to trans-disciplinary topics between fundamental science and engineering.

More information about this series at http://www.springer.com/series/8884

Michał Niełaczny · Barnat Wiesław
Tomasz Kapitaniak

Dynamics of the Unicycle

Modelling and Experimental Verification

 Springer

Michał Niełaczny
Division of Dynamics
Lodz University of Technology
Łódź, Poland

Tomasz Kapitaniak
Division of Dynamics
Lodz University of Technology
Łódź, Poland

Barnat Wiesław
Military University of Technology
Warsaw, Poland

ISSN 2191-530X   ISSN 2191-5318   (electronic)
SpringerBriefs in Applied Sciences and Technology
ISBN 978-3-319-95383-0   ISBN 978-3-319-95384-7   (eBook)
https://doi.org/10.1007/978-3-319-95384-7

Library of Congress Control Number: 2018946928

Printed on acid-free paper

This Springer imprint is published by the registered company Springer International Publishing AG part of Springer Nature
The registered company address is: Gewerbestrasse 11, 6330 Cham, Switzerland

Preface

As a qualified engineer and a sports enthusiast at the same time, the first author (ML) combines passion with science. The idea concerning the subject of this work appeared when the third author (TK) glimpsed ML unicycling along the university corridor.

The unicycle together with the rider forms a very complex system. It combines mechanics, biomechanics and control theory into a structure which impresses with both its simplicity and improbability. Even more amazing is that the majority of unicyclists do not have the faintest idea that riding such an artefact is, according to science, almost impossible. A similar situation is exemplified by bumblebees unaware of the fact that they should not be able to fly at all.

The present work is devoted to the problem of modelling and control of the 3-D dynamical system consisting of a single-wheel vehicle represented by a unicycle and a cyclist (to be precise—a unicyclist) riding it. The equations of motion were derived with the Boltzmann–Hamel equation in the matrix form, which is based on quasi-velocities and is usually scarcely used. The matrix form enables automatic generation of Hamel coefficients and eliminates all the difficulties associated with the determination of these quantities. The equations of motion were solved with Wolfram Mathematica. In order to make the unicyclist-imitating part of the model closer to reality, it was based on the main principles of biomechanics. The impact of a pneumatic tyre was investigated with the Pacejka Magic Formula scheme including the experimental determination of the stiffness coefficient.

The aim of control is to maintain the unicycle–unicyclist system in an unstable equilibrium around the given angular position. The control system based on the LQ Regulator was applied in Wolfram Mathematica. In order to examine the legitimacy of the model experimental validation, the 3-D motion capture using the OptiTrack Motive: Body software together with high-speed cameras was arranged. A description of the unicycle–unicyclist dynamical model, simulation results and experimental validation of the system are presented in the work.

This book is organized as follows: designs of different types of unicycles are described in Chap. 1, a model which represents the unicyclist riding the unicycle straight ahead is derived in Chap. 2, whereas numerical and experimental validation of the model is shown in Chap. 3. Finally, concluding remarks are drawn in Chap. 4.

We would like to acknowledge the helpful suggestions and discussions with Jarosław Strzałko, Juliusz Grabski and Jerzy Wojewoda.

Finally, we would like to acknowledge the help and support of our families and friends.

Łódź, Poland   Michał Niełaczny
Warsaw, Poland   Barnat Wiesław
Łódź, Poland   Tomasz Kapitaniak
April 2018

Contents

Notations

a_{ij}	Element of the matrix A
A	Matrix transforming generalized velocities into quasi-velocities
$\mathbb{A}$	State matrix
B	Inverse matrix to the matrix A
b_{nj}	Element of the matrix B
$\mathbb{B}$	Input matrix
CP	Contact point of the unicycle wheel and ground
C_r, C_l	Points, ends of each crank (left, right)
D	Derivative of the matrix A with respect to the vector q
d	Damping coefficient
E_{ju}	Coefficient in the Pacejka Magic Formula
F	Force in the Pacejka Magic Formula
g	Gravitational acceleration
g_j	Control gain of the j-th state variable
G	Three-dimensional matrix $(k \times k \times k)$ of Hamel symbols
H	Point, centre of mass of the wheel
H_r, H_l	Points, projection of H, on each leg motion plane
i	Number of the link
I_{ix}	Mass moment of inertia of the i-th link related to the axis x
I_{iy}	Mass moment of inertia of the i-th link related to the axis y
I_{iz}	Mass moment of inertia of the i-th link related to the axis z
I_i	Matrix of mass moments of inertia of the i-th link
$\mathbb{I}$	Control identity matrix
j	Next natural numbers, $j = (1, 2, 3 \ldots)$
k_t	Unicycle tyre stiffness coefficient
k_r	Unicycle rim stiffness coefficient
K_4, K_6	Points, each knee
$\mathbb{K}$	Law that minimizes the value of the cost
l_i	Auxiliary length of the i-th link
L_i	Length of the i-th link

m_i	Auxiliary mass of the i-th link
M_i	Mass of the i-th link
$\boldsymbol{M}_i$	Identity matrix of mass of the i-th link
O	Point, origin of the fixed frame $Oxyz$
P_4, P_6	Points, centre of mass of each pedal
q	Generalized coordinate
Q	External force
$\mathbb{Q}$	Matrix defines weights on states
r_i	Auxiliary radius of the i-th link
R_i	Radius of the i-th link
$\boldsymbol{R}_{\alpha i}$	Coordinates transformation matrix—rotation by the angle α_i around the axis z_{0i}
$\boldsymbol{R}_{\beta i}$	Coordinates transformation matrix—rotation by the angle β_i around the axis x_{0i}
$\boldsymbol{R}_{\gamma i}$	Coordinates transformation matrix—rotation by the angle γ_i around the axis z_i
$\boldsymbol{R}_i$	Coordinates transformation matrix of the i-th link from $\nabla\xi_i\eta_i\zeta_i$ to $Oxyz$
$\mathbb{R}$	Matrix defines weights on a control input in the cost function
S	Contact point of the saddle with the body
S_v	Offset in the Pacejka MF
t	Time
T	Kinetic energy
T^*	Kinetic energy expressed by quasi-velocities and generalized coordinates
u	Input quantity in the Pacejka MF
$\boldsymbol{u}$	Feedback control
v	Linear velocity
$\boldsymbol{v}$	Vector of linear velocities of the dimension $(k \times 1)$
V	Potential energy
w	Quasi-velocity
$\boldsymbol{w}$	Vector of quasi-velocities of the dimension $(k \times 1)$
$\boldsymbol{x}$	State variable
α_i	Angle of rotation around the axis z_{1i}
β_i	Angle of rotation around the axis x_{2i}
γ_i	Angle of rotation around the axis z_{3i}
$\left[\gamma^i_{nj}\right]$	Hamel coefficient
μ, χ, δ, ν	Angles to derive general coordinates directly related to the leg
Λ_{ij}	Coefficients for the leg motion, shorten notation
π	Number pi
ϖ	Correlation coefficient
ρ	Variable
ρ_i	Density of the i-th link
φ	Relative angle of rotation of the crank with respect to the frame
ψ	Quasi-coordinate

Ψ	Load related to the quasi-coordinate
ω	Angular velocity
$\boldsymbol{\omega}$	Vector of angular velocities of the dimension $(k \times 1)$
$Oxyz$	Main fixed inertial frame, attached in the point O
∇	Any point, the origin of the frame
$\nabla x'y'z'$	Moving non-inertial frame, parallel to $Oxyz$, attached in the point ∇
$\nabla x_{0i}y_{0i}z_{0i}$	Moving non-inertial frame, related to the i-th link, attached in the point ∇
$\nabla x_i y_i z_i$	Moving non-inertial frame, related to the i-th link, attached in the point ∇
$\nabla \xi_i \eta_i \zeta_i$	Moving non-inertial frame, pegged to the i-th link, attached in the point ∇

Chapter 1
Introduction

1.1 Unicycle—A One-Wheel Vehicle

A unicycle, which stands for a one-wheel vehicle, is a specific type of the bicycle equipped with one wheel only. The unicycle is considered the progeny of a penny-farthing, which is a bicycle with a large front driving wheel and a small rolling rear one [1]. A penny-farthing and a unicycle are shown in Figs. 1.1 and 1.2, respectively. The unicycle was built in the late nineteenth century after the removal of the rear wheel and the frame from the penny-farthing [4]. The unicycle consists of a remarkably fewer constituents than the regular bicycle. It may seem that its parts resemble the bicycle counterparts, but the similarity is purely visual.

A fixed gear is the main feature of the unicycle, which means that cranks are integral with the wheel and, thus, cannot be separated from it. Therefore, the rotation of cranks controls the rotation of the wheel directly. When pedalling is not involved, riding is impossible. The specificity of this mechanism enables the user to ride backwards or stand up, but only one gear is available.

Currently, the most popular are unicycles with 20-inch (20") wheels. The advantages of them are as follows:

- compact dimensions,
- lightweight structure,
- they can reach the speed of about 8 km/h,
- they cannot be more unsophisticated mechanically,
- uncomplicated servicing due to a modest number of parts needed,
- unicyclist's position while riding follows the physiological/natural upright position,
- they arouse the interest of pedestrians,
- law regulations concerning unicycles are imprecise.

Acquiring the necessary skills to ride a unicycle is slightly more sophisticated and time-consuming than learning to ride a regular bike. The gravity centre of an

© The Author(s) 2019
M. Niełaczny et al., *Dynamics of the Unicycle*,
SpringerBriefs in Applied Sciences and Technology,
https://doi.org/10.1007/978-3-319-95384-7_1

Fig. 1.1 Penny-farthing—
popular at the end of the
nineteenth century in the
Skoda Auto Museum, Czech
Republic, [2]

Fig. 1.2 QU-AX Luxus
20—currently the most
common unicycle, [3].
Courtesy of QU-AX GmbH

Fig. 1.3 Comparison of
location of the centre of
gravity between the
pedestrian and the unicyclist

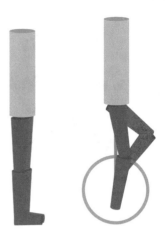

average person while standing is situated around his (or her) belly button, as shown in Fig. 1.3 (the centre of gravity is situated shortly above the boundary between the grey and blue segments), [5]. A unicyclist is able to maintain easily the upright position as well. However, his gravity centre is higher (a larger distance from the ground) than in the case of a regular pedestrian, even when its lowest position is taken into consideration. There is no doubt that it makes riding slightly more demanding.

An ordinary man in the street is usually inclined to be lazy by nature, therefore becomes discouraged and abandons further pursuit after a few unsuccessful attempts. The first author of this book is an avid user of unicycles and opposes the stereotypes according to which the falls are extremely painful and lead to serious body injuries. The second factor which makes riding a unicycle more strenuous and demanding than using a traditional bike is only one point of support. Additionally, it is considerably smaller than the sole surface of two feet. For the foregoing reason, the balance needs to be maintained in two planes simultaneously. Sustaining the balance in the plane which is perpendicular to the riding direction (left/right) involves S-style riding (using the centrifugal force) and hip balancing in order to achieve and maintain the centre of gravity above the fulcrum of the wheel. In contrast, keeping the balance in the plane parallel to the direction of our locomotion (forward/backward) consists in accelerating and slowing down the drive wheel so that the gravity centre oscillates above the fulcrum.

1.2   Types of Unicycles

Unicycles can be divided into several types as regards either the size of the driving wheel or their purpose. Both of these criteria are closely connected, because an appropriate wheel size is assigned to a particular discipline in order to provide the user with the most comfortable riding experience.

Fig. 1.4 Different sizes of unicycles

Standard wheel sizes are as follows: 20", 24", 26", 27.5", 29" and 36" as shown in Fig. 1.4 [4]. However, wheels of smaller sizes, for instance 12", 16" or 18", exist as well—they are dedicated to children. Wheels larger than 36" are not in use. The limited length of human legs does not allow us to use them. Unicycles are used for a variety of sports like regular two-wheel bicycles. Unicycles designed for extreme sports, e.g., street or flatland trials, can be used to jump and perform complex manoeuvres, just like with skateboards (Fig. 1.5). Such a unicycle has a small wheel, typically 20", its tyre is thick and its construction—reinforced. The pressure in its tyre is relatively low and reaches at most 2 *bar*. These factors allow the user to make much more impressive jumps and experience the effect of cushion landing. While performing the tricks, the unicyclist usually stands on the pedals, grabbing the saddle in his/her hand. The *downhill* style involves using 24", 26" and 27.5" unicycles. Their contraction is similar to that of *street unicycles*, although their wheels are larger. *Downhill* unicycles are additionally equipped with a brake. There are two particular types of brakes used in unicycling:

- hydraulic rim brake,
- hydraulic disc brake (disc mounted as a chaining onto the right crank).

Fig. 1.5 Trial unicycling [6]

Fig. 1.6 Road unicycling [6]

When riding on an even surface, the fulcrum of the wheel is situated directly beneath it. Due to having just one wheel at their disposal, the unicyclist is able to maintain the upright position while riding down the hill, but the fulcrum is moved slightly backwards. Road riding or long-distance riding are immensely popular, too. The larger the wheel and the shorter the crank, the faster the ride. An example of the road unicycle is shown in Fig. 1.6. Like *downhill* unicycles, such road unicycles are equipped with a brake. Moreover, they comprise a handlebar called a *T-Bar* and a geared hub—*Schlumpf* [6]. Additionally, the unicycles designed for long-distance riding have carriers with panniers for storing the luggage.

The aim of the above description is to present various types of unicycles. However, several variations of these vehicles exist, and the most popular one is shown in Fig. 1.4. In the further part of the study, this unicycle will be subject to analysis.

1.3 Unicycle—Technical Aspects

From the mechanical aspect, the unicycle–unicyclist system can be treated as a moving (with the velocity v) double-inverted spherical pendulum [7–10]. To generalize, the unicyclist's trunk represents the first link (upper pendulum bob), whereas the frame of the unicycle forms the second link (lower pendulum bob) as shown in Fig. 1.7. Assuming that the wheel is one of pendulum's links as well, the model acquires the form of a triple spherical pendulum [11, 12]. Such an attitude to the unicycle–unicyclist system perfectly corresponds to reality—standing at a particular point without balancing is impossible [13–16].

When riding a unicycle, the upright position is the initial one. Let us investigate the case in which the unicyclist is beginning to loose his/her balance. By means of a measuring element represented by the membranous labyrinth, he/she feels that he/she

Fig. 1.7 Double-inverted
pendulum model of the
unicycle–unicyclist system

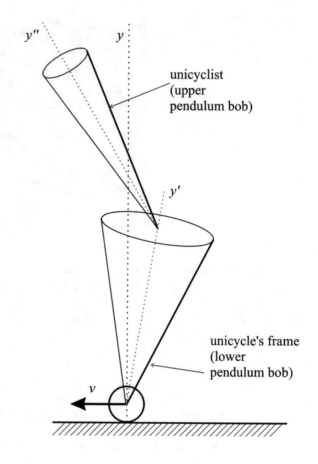

y'' y

unicyclist
(upper
pendulum bob)

y'

unicycle's frame
(lower
pendulum bob)

v

swings from the position of unstable equilibrium. Such a deflection is treated as a control error. Consequently, the control unit, in this case the unicyclist's brain, sends the signals to the appropriate extremities (actuators). Eventually, the rider balances with his whole body and returns to the upright position. To summarise, the unicycle together with the cyclist who has mounted it can be treated as a follow-up control system in the control aspect (Fig. 1.8).

In most cases, when scientists use the term *unicycles*, they mean non-holonomic wheeled mobile robots equipped with a steering wheel (unicycle) and two independent drive wheels [1–3], which come in handy when facing the track control problems.

The unicycle, which is an object of these investigations, is simplified to a wheel with an inverted pendulum, sometimes even double, attached to its axle. The second link mentioned is supposed to simulate the unicyclist [4–12, 17]. Unicyclists are usually treated cursorily and conventionally. Moreover, their lower extremities are usually skipped or reduced to the unicycle frame at best. In fact, the role of legs is essential, as due to their presence the imbalance of the wheel caused by the cranks and

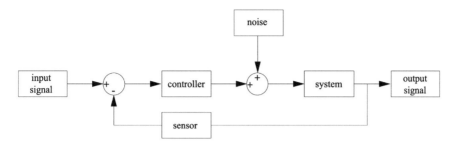

Fig. 1.8 Follow-up control system

pedals is increased. The imbalance is responsible for the limited speed the unicyclist is able to achieve during the ride. In other words, it is the legs that make unicycle riding such a challenge. Just in a few cases [11–15], the impact of lower extremities is taken into consideration, but still cursorily, superficially and without an application of biomechanical rules to them.

As a consequence of their design, such unicycle-like vehicles occasionally occur, e.g., in the control theory. It is indeed a challenge to control an unstable device with only one point of contact with the ground. In fact, what can be investigated, studied and tested involving the unicycle-like vehicles is more commonplace than attempts to build a biomechanical model reflecting the behaviour of the unicyclist during the ride.

Additionally and unfortunately, such a unicycle-like robot is considered to be a biomechanical system only in terms of using the internal world representation, containing the description of emotion, instinct and intuition mechanisms. Researches are focused on intelligent control methods, intuition and biological instinct, which are essential in imitating human behaviours and actions. In other words, the research is conducted on soft computing algorithms to examine the intelligent control on the example of a unicycle robot. Moreover, none of the experiments includes a comparison of the results with a real object (real unicycle–unicycling system), i.e., the motion capture. Like in the case of validation based on a real object, none of the models was analysed in terms of a pneumatic tyre, although it is only the tyre—among all parts of the unicyclist–unicycle system—that maintains contact with the ground.

In all likelihood, this is due to the fact that the scientists who tackle this issue are not active riders themselves. They aim to investigate other phenomena on the basis of this specific system. Actual riders, the first author of this book amongst them, perceive the unicycle–unicyclist system model as the most valuable, because unicycling itself can be significantly improved thanks to it.

References

1. Limebeer, D., & Sharp, R. (2006). Bicycles, motorcycles, and models. *IEEE Control Systems*, *26*, 34–61.
2. https://commons.wikimedia.org/wiki/File:Ordinary_bicycle01.jpg. Retrieved 4, 2018.
3. https://www.qu-ax.de/en/. Retrieved 1, 2017.
4. Cossalter, V. (2006). *Motorcycle dynamics*. Morrisville: Lulu Press.
5. Clauser, C. E., McConville, J. T., & Young, J. W. (1971). Weight, volume, and center of mass of segments of the human body. *Journal of Occupational and Environmental Medicine*, *13*, 270–280.
6. http://www.krisholm.com/en/. Retrieved 3, 2015. (Figure 1.5 presents Max Schulz (photo by George Smith). Scott Wilton is shown in Figure 1.6 (photo by Warren Howell). Courtesy of Kris Holm).
7. Wendlandt, J. (1995). *Pattern evocation and energy-momentum integration of the double spherical pendulum*. M.A. Thesis, University of California.
8. Marsden, J., & Scheurle, J. (1993). *Lagrangian reduction and the double spherical pendulum*. Basel: Birkhauser Verlag.
9. Marsden, J. (1995). Visualization of orbits and pattern evocation for the double spherical pendulum. In *Proceedings of the ICIAM Conference, Hamburg*.
10. Cheng, F., Zhong, G., Li, Y., & Xu, Z. (1996). Fuzzy control of a double-inverted pendulum. *Fuzzy Sets and Systems*, *79*, 315–321.
11. Yeung, S. (1995). *The triple spherical pendulum*. Technical Report, California Institute of Technology, Division of Chemistry and Chemical Engineering.
12. Hoshino, T., Kawai, H., & Furuta, K. (2009). Methoden und Anwendugen der Steuerungs-, Regelungs- und Informationstechnik, *48*, 577–587.
13. Furut, K., Ochiai, T., & Ono, N. (1984). Attitude control of a triple inverted pendulum. *International Journal of Control*, *36*, 1351–1365.
14. Kajita, S., Kanehiro, F., Kando, K., Yokoi, K., & Hirukawa, H. (2001). The 3D linear inverted pendulum mode: A simple modeling for a biped walking pattern generation. In *Proceedings of the 2001 IEEE/RSJ International Conference on Intelligent Robots and Systems, IEEE*(pp. 239–246).
15. Rong, Y. (2000) *Geometric techniques for control of a 2-DOF spherical inverted pendulum*. Ph.D. Thesis, University of Science and Technology, Hong Kong.
16. Zhong, W., & Rock, H. (2001). Energy and passivity based control of the double inverted pendulum on a cart. In *Proceedings of the 2001 IEEE International Conference on Control Applications, IEEE* (pp. 896–900).
17. http://sillycycle.com/crab.html. Retrieved 5, 2014.

Chapter 2
Model of the Unicycle-Unicyclist System

At this point, the "everyday" typical ride performed by an experienced cyclist is pondered over. The case of a unicyclist riding on an even surface from the point A to B is taken under consideration. While covering a given distance, the user maintains the straight, vertical position on the saddle. The unicyclist keeps usually his/her hands on the saddle front bumper or positions the arms laterally. Such an alteration of the rider's position is helpful in three particular cases, namely:

- while learning to ride a unicycle,
- while performing a trick, for instance jumping, or riding along rails,
- while trying to overcome the imbalance caused by an impact against the uneven ground.

In accordance with the given information, the influence of upper extremities is reduced while riding, and therefore, they can be treated as the whole together with the unicyclist's trunk (as a combination of masses).

Figure 2.1 depicts a model of the system which includes approximate shapes of particular links. As we delve into the work slightly further, the wire model drawing presented in Fig. 2.2 would supersede the previously mentioned one for the better clarity of considerations. In this figure, the wires representing the links are covered with the axis η of the link.

To simplify the investigations of the unicycle-unicyclist system, the following assumptions are introduced:

- **Wheel**—consists of a rim (a circular hoop) with a tyre attached to it. It is in the centre of the rim where the hub (cylinder) is located. It is connected permanently to cranks (cuboid). A pedal (cuboid) is attached to the end of each crank. The generalized coordinates directly related to the wheel are presented below:

© The Author(s) 2019
M. Niełaczny et al., *Dynamics of the Unicycle*,
SpringerBriefs in Applied Sciences and Technology,
https://doi.org/10.1007/978-3-319-95384-7_2

Fig. 2.1 Model of the
system prepared with
Wolfram Mathematica^{TM}

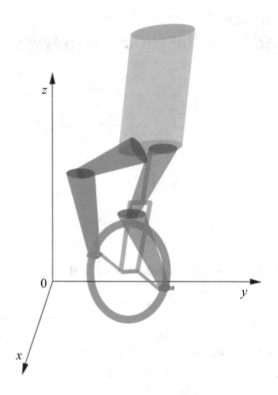

$$q_1 = \begin{bmatrix} x_1 \\ y_1 \\ z_1 \\ \alpha_1 \\ \beta_1 \\ \gamma_1 \end{bmatrix}. \tag{2.1}$$

- **Frame**—is attached to the hub with a rotary knot. At the extreme point of the frame (the fork consisting of elliptical and cylindrical shields), there is a saddle (a torus segment). The generalized coordinates directly related to the frame are as follows:

$$q_f = [\alpha_2]. \tag{2.2}$$

- **Body**—is connected to the saddle with a spherical knot. During the motion under consideration, the unicyclist keeps their arms laterally or places them on the front saddle bumper. Therefore, the cyclist's trunk together with his/her arms is modelled as an elliptical cylinder. The generalized coordinates straightly related to the body are presented as:

Fig. 2.2 Wire model of the system prepared with *Wolfram Mathematica*^{TM}

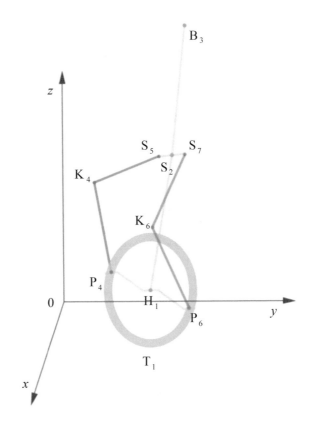

$$q_3 = \begin{bmatrix} \alpha_3 \\ \beta_3 \end{bmatrix}.$$ (2.3)

- **Tibia** (right and left)—each tibia is attached to the right or left pedal with a rotary knot. The model of tibia resembles a reduced/cut cone having its centre of gravity at $\frac{3}{4}$ of its height. The generalized coordinates related directly to the right and left tibia are as follows:

$$q_4 = [\alpha_4],$$ (2.4)

$$q_6 = [\alpha_6].$$ (2.5)

- **Thigh** (right and left)—each thigh is not only connected to the right or left tibia, respectively, but also to the body by means of rotary knots. Similarly to the tibia, the model of thigh resembles a cut cone with the gravity centre at approximately $\frac{3}{4}$ of its height. The generalized coordinates directly related to the right and left thigh are expressed in the form:

$$q_5 = [\alpha_5], \tag{2.6}$$

$$q_7 = [\alpha_7]. \tag{2.7}$$

To conclude, the model comprises 7 particular components/bodies.

Highlighting the presence of unicyclist's legs is the main advantage of such a model, as they contribute to increasing the imbalance of the wheel caused by cranks and pedals and, therefore, their role seems crucial. Lower extremities are usually omitted when models of such systems [1–5] are generated, although they are indeed the main factor which makes unicycling so tough.

The unicyclist's leg used in the model consists of a thigh and a tibia. The foot is omitted due to its specific and complex motion in the rotational cycle which does not have any significant impact on conventional riding, thus can be neglected. Therefore, the pedal axle is covered with an ankle. Thanks to such a construction, the lower extremity can be treated as a crank mechanism and the leg's position can be defined clearly by the angles of rotation of the driving wheel γ_1 and with the frame rotation angle around the wheel α_2, according to Eqs. (2.8)–(2.15), as in the example of the right leg. The relative angle of rotation of the right crank in relation to the frame is:

$$\varphi_4 = \gamma_1 - \alpha_2. \tag{2.8}$$

The distance between the pedal and the hip is:

$$P_4 S_4 = \sqrt{[H_4 S_4 - H_4 P_4 \cos \varphi_4]^2 + [H_4 P_4 \sin \varphi_4]^2}. \tag{2.9}$$

The angles essential to derive the general coordinates directly related to the leg—α_4 and α_5:

$$\mu_4 = \arctan \frac{H_4 P_4 \sin \varphi_4}{H_4 S_4 - H_4 P_4 \cos \varphi_4}, \tag{2.10}$$

$$\chi_4 = \arccos \frac{(P_4 S_4)^2 + (P_4 K_4)^2 - (K_4 S_4)^2}{2 \cdot P_4 K_4 \cdot P_4 S_4}, \tag{2.11}$$

$$\delta_4 = \arccos \frac{(P_4 K_4)^2 + (K_4 S_4)^2 - (P_4 S_4)^2}{2 \cdot P_4 K_4 \cdot K_4 S_4}, \tag{2.12}$$

$$\nu_4 = \chi_4 - \mu_4, \tag{2.13}$$

$$\alpha_4 = 2\pi - (\varphi_4 - \nu_4), \tag{2.14}$$

$$\alpha_5 = 2\pi - (\pi - \delta_4). \tag{2.15}$$

Equations (2.16)–(2.18) show formulas for location of the right knee according to Appendix A, Eq. (A.6):

$$\boldsymbol{r}_{K4} = \begin{bmatrix} 0 \\ L_4 \\ 0 \end{bmatrix},\tag{2.16}$$

$$\boldsymbol{R}_4 = \boldsymbol{R}_1 \boldsymbol{R}_{\alpha_4} = \boldsymbol{R}_{\alpha_1} \boldsymbol{R}_{\beta_1} \boldsymbol{R}_{\gamma_1} \boldsymbol{R}_{\alpha_4},\tag{2.17}$$

$$\boldsymbol{r}_{OK_4} = \boldsymbol{r}_{OP_4} + \boldsymbol{R}_4 \boldsymbol{r}_{K4}.\tag{2.18}$$

The relative angle of rotation of the left crank with respect to the frame equals:

$$\varphi_6 = \varphi_4 + \pi.\tag{2.19}$$

The procedure used for the left leg is analogical as for right one.

2.1 Parameters of the Model

2.1.1 Unicycle Parameters

For the purpose of simulations, each and every parameter of the unicycle parts which was used during the investigations was derived from the manufacturers' specifications [2, 6, 7].

2.1.2 Unicyclist's Parameters and Biomechanics

In order to make the unicyclist's part of the model closer to reality, it covers the following aspects based on the main principles of biomechanics:

- presence of the synovial fluid in joints,
- natural shape of modelled body parts,
- location of the centre of mass.

Three-dimensional body scans were made so that the length of unicyclist's limbs could be measured. Additionally, the circular symmetry markers were used. They were arranged in the room to create an XYZ frame. Photographs were taken from various perspectives and then uploaded to the *GOM TRITOP* software [8]. This software is used for analysing high-resolution images and calculating 3-D coordinates. The *GOM TRITOP* software could serve as a tool for measuring and inspecting not only minor components but also considerable elements such as wind power units or even parts of ships. Its basic functions include capturing circular symmetry markers and processing them to create 3-D point clouds as well as monitoring the quality of the image, its mapping and 3-D views. After the 3-D coordinates were determined, the whole measurement was mathematically transformed into the coordinate system of the component. Finally, the lines linking the markers in each of the directions were generated and compared to the theoretical values [9].

1. The horizontal circumference measured at the thickest part of the calf.
2. The horizontal circumference measured at the thickest part of the thigh.
3. The horizontal circumference measured at the largest trunk narrowing.
4. The horizontal chest circumference measured at the breast height.

The locations of the centre of mass and the mass values attributed to particular body parts were selected on the basis of the theoretical values [9, 10]. The quantity/value of the synovial fluids present in joints such as the hip or knee is based on the theoretical values [11].

As mentioned in this chapter, unicyclist's arms were integrated with the trunk and modelled as an elliptic cylinder (Fig. 2.1). In contrast, particular parts of cyclist's legs were created to resemble cut cones, so that the mass distribution corresponds to the theoretical mass centre. Figure 2.10 depicts an example of the modelled unicyclist's thigh.

According to biomechanics, the mass centre of calves and thighs should be located at 43.3% of their height [9, 10]:

$$M_5 = \frac{1}{3}\rho_5\pi \left(R_5^2 L_5 - r_5^2 l_5\right), \quad m_5 = \frac{1}{3}\rho\pi r_5^2 l_5, \tag{2.20}$$

$$0.433 L_5 = \frac{L_5 \left(R_5^2 + 2r_5 R_5 + 3r_5^2\right)}{4\left(R_5^2 + r_5 R_5 + r_5^2\right)}, \tag{2.21}$$

$$\frac{l_5 + L_5}{R_5} = \frac{l_5}{r_5}. \tag{2.22}$$

The moments of inertia ascribed to the C_5 centre of mass are as follows:

$$I_{5x} = I_{5y} = \frac{3}{20}\left((M_5 + m_5)\left(R_5^2 + 4\left(L_5 + l_5\right)^2\right) - m_5\left(r_5^2 + 4l_5^2\right)\right), \tag{2.23}$$

$$I_{5z} = \frac{3}{10}\left(R_5^2\left(M_5 + m_5\right) - m_5 r_5^2\right). \tag{2.24}$$

The data derived from Tables 2.1 and 2.2 and Eqs. (2.20)–(2.22) allow one to calculate the radius r_5 as well as the mass moments of inertia ascribed to the cut cone of specific features. Such modelled leg links were used in numerical simulations.

2.2   Tyre Modelling

A tyre plays a vital role in operation of the unicycle. At first, it influences the riding comfort and adhesion to the ground. Secondly, the adequate pressure inside the tyre is helpful when trying to maintain the balance. The latter factor comes useful in

Table 2.1 Measurements of unicyclist's limbs

Markers	x, mm	y, mm	z, mm	Resultant, mm	Theoretical, mm
Head–shoulder	44.163	322.657	143.154	355.740	324
Shoulder–hip	46.063	451.708	20.268	454.503	460
Hip–knee	271.855	321.164	14.473	421.024	441
Knee–pedal	136.009	462.655	55.916	485.464	495

Table 2.2 Circumferences of unicyclist's extremities

Body part	Experimental, mm	Theoretical, mm
1	355	367
2	506	519
3	801	872
4	921	948

extreme riding, while performing tricks and evolutions, Fig. 1.5. Such a value of the pressure is definitely lower in comparison with its values used in ordinary riding, Sect. 1.3.

2.2.1 Tyre Stiffness Coefficient

An *Instron 4485 Testing Machine* is a universal, electromechanical, static testing system which performs tensile and compression testing, as well as other mechanics-related analyses which meet industrial standards. Its frames are able to comply with the requirements of both the extended travel and wide test space. The leading advantage of such a testing machine is the appropriate stiffness of the frame and an easy access to the testing area (Fig. 2.11). When examining the elements of a more considerable size, the frames characterised by the lower base height allow the operator to stand closer to grips and fixtures [12].

The tyre stiffness coefficient k_t, indispensable to the Pacejka Magic Formula (Sect. 2.2.2), was evaluated with an *Instron 4485* testing machine. The maximum vertical load applied to a wheel has the value of 75 dN, which is equivalent to the weight of an average unicyclist. The pressure inside the tyre achieved the value of 6 bar, which is recommended for this particular type of the tyre. The deflection calculated in [mm] was checked, as well as the wheel stiffness with and without the tyre, on the basis of its load and deflections. Three trials were performed in each case. As a result of the experiment, the following stiffness coefficients were estimated:

- $k_t = 119047.6$ N/m,
- $k_r = 1086956.5$ N/m.

The outcomes of the experiments presented above indicate that the tyre impact on the unicycle operation is insignificant. In the further part of work, the wheel will be regarded as a rigid body. Nevertheless, the tyre model (Sect. 2.2.2) based on the examination was attached to the system to confirm this assumption. Next, the results are compared in Chap. 3.

2.2.2   Pacejka Magic Formula

Diagonal tyres enjoy the greatest popularity among cyclists (as well as unicyclists). In spite of this fact, radial tyres are in use likewise, for instance *MAXXIS Radiale* [13]. According to this information and a proposal from Sect. 2.2.1, the Pacejka Model—*Magic Formula* involving a radial tyre was used to simulate the unicycle tyre. Figure 2.14 depicts a scheme of the Magic Formula [14].

The fact that slippage is ignored in the consideration (Sect. 2.4) follows from limited velocity values and the investigation of a classical, straight ride on an even surface. Actually, the Magic Formula can be simplified to three inputs only.

In terms of the tyre model, the Pacejka Magic Formula is a system of equations which approximates test curves. The general equation has the form of:

$$F_u = E_{3u} \sin \left(E_{2u} \arctan \left(E_{1u} u - E_{4u} \left(\arctan \left(E_{1u} u \right) \right) \right) \right) + S_y. \qquad (2.25)$$

With Eq. (2.25), longitudinal or lateral tyre forces as well as the vertical moment can be calculated. The input quantity u can take the values of a longitudinal slip or a camber angle. Both the experimental stiffness coefficient (Sect. 2.2.1) and the camber angle from the numerical simulations were used in the calculations. The remaining coefficients and factors are based on the theoretical values. Chapter 3 comprises a comparison of the systems distinguished by the presence or absence of the pneumatic tyre. On this basis, the assumption on the insignificance of the tyre impact (from Sect. 2.2.1) is confirmed.

2.3   Boltzmann–Hamel Equations

In most cases, the Lagrange equations of the second kind are used to derivative the motion equations of a holonomic system [16]:

$$\frac{d}{dt} \left(\frac{\partial T}{\partial \dot{q}} \right) - \frac{\partial T}{\partial q} + \frac{\partial V}{\partial q} = Q. \qquad (2.26)$$

They allow one to obtain the equations of motion on the basis of the generalized coordinates. Quasi-velocities are of great convenience when describing the movement in variable configurations for holonomic and non-holonomic systems.

An introduction of quasi-velocities into the system motion characterization is favourable as long as their use provides a compact notation of kinetic energy and motion equations, e.g., on examination/investigation of the systems which contain relative movement producing elements. The Boltzmann–Hamel equations are scarcely used, due to a complicated formula containing the Hamel coefficients (Eq. (2.28)) as well as sophisticated relations for determining these coefficients [17–21]. The typical form of the Boltzmann–Hamel equations dedicated to the system with a number of generalized coordinates equal to k presents as follows:

$$\frac{d}{dt}\left(\frac{\partial T^*}{\partial w_n}\right) - \frac{\partial T^*}{\partial \psi_n} + \sum_{m=1}^{m=k}\sum_{l=1}^{l=k}\sum_{i=1}^{i=k}\sum_{j=1}^{j=k} b_{li}b_{mj}\left(\frac{\partial a_{im}}{\partial q_l} - \frac{\partial a_{il}}{\partial q_m}\right)\frac{\partial T^*}{\partial w_i}w_i = \Psi_n,$$

$$(n = 1,\ldots k). \quad (2.27)$$

Introducing the Hamel coefficients γ_{nj}^i defined as:

$$\gamma_{nj}^i = \sum_{m=1}^{m=k}\sum_{l=1}^{l=k} b_{ln}b_{mj}\left(\frac{\partial a_{im}}{\partial q_l} - \frac{\partial a_{il}}{\partial q_m}\right), \quad (2.28)$$

we obtain a simple form of the Boltzmann–Hamel equations:

$$\frac{d}{dt}\left(\frac{\partial T^*}{\partial w_n}\right) - \frac{\partial T^*}{\partial \psi_n} + \sum_{i=1}^{i=k}\sum_{j=1}^{j=k} \gamma_{nj}^i \frac{\partial T^*}{\partial w_i}w_j = \Psi_n. \quad (2.29)$$

The matrix form [22] of the Boltzmann–Hamel equations (Eq. (2.30)) enables one to automate the generation of the Hamel coefficients and eliminate all the difficulties associated with determination of these quantities.

$$\frac{d}{dt}\left(\frac{\partial T^*}{\partial w}\right) + \left(G^T w\right)\frac{\partial T^*}{\partial w} - B^T\frac{\partial T^*}{\partial q} = B^T\left(f - \frac{\partial V}{\partial q}\right), \quad (2.30)$$

where:

$$\left(B = A^{-1}\right), \quad (2.31)$$

$$D = \left[\frac{\partial A}{\partial q}\right], \quad D_i = \frac{\partial a_i}{\partial q}, \quad (2.32)$$

$$G^i = \left[\gamma_{nj}^i\right] = B^T\left(D_i - D_i^T\right)B, \quad (n, j = 1, \ldots, k). \quad (2.33)$$

The quasi-coordinates ψ_n are integrals of the quasi-velocities w_n, which are defined as linear functions of the generalized velocities $\dot{q}_n$ in the noninertial frame:

$$w_n = \sum_{j=1}^{j=k} a_{nj} \dot{g}_j, \qquad (n = 1, \ldots k). \tag{2.34}$$

The quasi-velocities are independent/autonomous and their amount is equal to that of the generalized velocities k. The introduction of a vector of quasi-velocities presents as follows:

$$w = A\,(q)\,\dot{q}. \tag{2.35}$$

The generalized velocities defined by the quasi-velocities occur in the relation opposite to Eq. (2.34):

$$\dot{q}_n = \sum_{j=1}^{j=k} b_{nj} w_j, \qquad (n = 1, \ldots k). \tag{2.36}$$

The given relation could be expressed as:

$$\dot{q} = B\,(q)\,w, \tag{2.37}$$

The equations of motion ascribed to the unicycle-unicyclist system dynamics (Eq. (2.68)), based on the matrix form of the Boltzmann–Hamel equation (Eq. (2.30)), were solved with *Wolfram Mathematica^{TM}*.

2.4   Energy of the System

The kinetic energy of the unicycle-unicyclist system is equivalent to a sum of the linear and circular motion of each of seven links (this chapter) [16, 22, 23]. It is expressed by the following formula:

$$T^* = \frac{1}{2} \sum_{i=1}^{7} v_i^T M_i v_i + \frac{1}{2} \sum_{i=1}^{7} \omega_i^T I_i \omega_i \qquad i = (1, \ldots, 7), \tag{2.38}$$

where:

$$v_i = \begin{bmatrix} v_{i\xi} \\ v_{i\eta} \\ v_{i\zeta} \end{bmatrix}, \tag{2.39}$$

$$\boldsymbol{M}_i = \begin{bmatrix} M_i & 0 & 0 \\ 0 & M_i & 0 \\ 0 & 0 & M_i \end{bmatrix}, \tag{2.40}$$

$$\boldsymbol{\omega}_i = \begin{bmatrix} \omega_{i\xi} \\ \omega_{i\eta} \\ \omega_{i\zeta} \end{bmatrix}, \tag{2.41}$$

$$\boldsymbol{I}_i = \begin{bmatrix} I_{i\xi} & I_{i\xi\eta} & I_{i\xi\zeta} \\ I_{i\xi\eta} & I_{i\eta} & I_{i\eta\zeta} \\ I_{i\xi\zeta} & I_{i\eta\zeta} & I_{i\zeta} \end{bmatrix}. \tag{2.42}$$

The linear velocity assigned to a particular point of the rigid body, in the vector notation within the frame $\nabla \xi_i \eta_i \zeta_i$, can be defined as:

$$\boldsymbol{v}_i = \boldsymbol{v}_0 + \boldsymbol{\omega}_i \times \boldsymbol{r}_i \tag{2.43}$$

where:

$$\boldsymbol{v}_i = \begin{bmatrix} v_{\xi i} & v_{\eta i} & v_{\zeta i} \end{bmatrix}, \tag{2.44}$$
$$\boldsymbol{v}_0 = \begin{bmatrix} v_{\xi 0} & v_{\eta 0} & v_{\zeta 0} \end{bmatrix}, \tag{2.45}$$
$$\boldsymbol{\omega}_i = \begin{bmatrix} \omega_{\xi i} & \omega_{\eta i} & \omega_{\zeta i} \end{bmatrix}, \tag{2.46}$$
$$\boldsymbol{r}_i = \begin{bmatrix} r_{\xi i} & r_{\eta i} & r_{\zeta i} \end{bmatrix}. \tag{2.47}$$

On the basis of Eq. (2.43)–(2.47), it is obtained:

$$\boldsymbol{v}_i = \begin{bmatrix} v_{\xi 0} & v_{\eta 0} & v_{\zeta 0} \end{bmatrix} + \begin{bmatrix} \omega_{\eta i} r_{\zeta i} - \omega_{\zeta i} r_{\eta i}, & \omega_{\zeta i} r_{\xi i} - \omega_{\xi i} r_{\zeta i}, & \omega_{\xi i} r_{\eta i} - \omega_{\eta i} r_{\xi i} \end{bmatrix}, \tag{2.48}$$

which can be expressed in a more convenient way as:

$$\boldsymbol{v}_i = \begin{bmatrix} v_{\xi 0} \\ v_{\eta 0} \\ v_{\zeta 0} \end{bmatrix} + \begin{bmatrix} \omega_{\eta i} r_{\zeta i} - \omega_{\zeta i} r_{\eta i} \\ \omega_{\zeta i} r_{\xi i} - \omega_{\xi i} r_{\zeta i} \\ \omega_{\xi i} r_{\eta i} - \omega_{\eta i} r_{\xi i} \end{bmatrix}. \tag{2.49}$$

This linear velocity v_i can be also presented in its matrix representation:

$$\boldsymbol{v}_i = \boldsymbol{v}_0 + \boldsymbol{\Omega}_i \boldsymbol{r}_i \tag{2.50}$$

where:

$$\boldsymbol{v}_i = \begin{bmatrix} v_{\xi 0} \\ v_{\eta 0} \\ v_{\zeta 0} \end{bmatrix}, \tag{2.51}$$

Table 2.3 Link number substitutions

i	Link
1	Wheel
2	Frame
3	Body
4	Right tibia
5	Right thigh
6	Left tibia
7	Left thigh

$$r_i = \begin{bmatrix} r_{\xi i} \\ r_{\eta i} \\ r_{\zeta i} \end{bmatrix}, \tag{2.52}$$

$$\boldsymbol{\Omega}_i = \boldsymbol{R}_i^T \dot{\boldsymbol{R}}_i, \quad \boldsymbol{R}_i = \boldsymbol{R}_{\alpha i} \boldsymbol{R}_{\beta i} \boldsymbol{R}_{\gamma i} \tag{2.53}$$

Knowing that the $\boldsymbol{\omega}_i \times \boldsymbol{r}_i$ and $\boldsymbol{\Omega}_i \boldsymbol{r}_i$ values are equal Eqs. (2.43), (2.50), it can be deduced that $\boldsymbol{\Omega}_i$ is a skew-symmetric matrix:

$$\boldsymbol{\Omega}_i = \begin{bmatrix} 0 & -\omega_{\zeta i} & \omega_{\eta i} \\ \omega_{\zeta i} & 0 & -\omega_{\xi i} \\ -\omega_{\eta i} & \omega_{\xi i} & 0 \end{bmatrix}. \tag{2.54}$$

The derivation of angular and linear velocities describing each of the links that the unicycle-unicyclist system consists of, which are substituted to Eq. (2.38), is represented by an example of the unicycle wheel and frame. The procedure is analogous for other links (Table 2.3).

The linear velocity of the wheel in the $H\xi_1\eta_1\zeta_1$ frame, where $v_0 = 0$ and the wheel radius is expressed in the frame $Hx_1y_1z_1$, takes the form:

$$v_1 = \boldsymbol{\Omega}_1 \boldsymbol{R}_{\gamma 1} r_1, \tag{2.55}$$

where:

$$\boldsymbol{\Omega}_1 = \boldsymbol{R}_1^T \dot{\boldsymbol{R}}_1, \quad \boldsymbol{R}_1 = \boldsymbol{R}_{\alpha 1} \boldsymbol{R}_{\beta 1} \boldsymbol{R}_{\gamma 1}. \tag{2.56}$$

The angular wheel velocity in the $H\xi_w\eta_w\zeta_w$ frame is as follows:

$$\boldsymbol{\omega}_1 = \begin{bmatrix} \omega_{1\alpha} \\ \omega_{1\beta} \\ \omega_{1\gamma} \end{bmatrix}. \tag{2.57}$$

The quasi-velocities of the wheel resulting from its circular motion (Fig. 2.3) are expressed in the form:

$$\boldsymbol{w}_1 = \boldsymbol{R}_{\alpha 1}\boldsymbol{\omega}_1 = \begin{bmatrix} w_{1\alpha} \\ w_{1\beta} \\ w_{1\gamma} \end{bmatrix}, \qquad (2.58)$$

When the motion of a sharp-edged disc, rolling on a plane, horizontal surface (slippage is omitted, Fig. 2.2) is considered, the non-holonomic constraints defining the wheel model, Fig. 2.2, are presented as in relation [22]:

$$\boldsymbol{w}_1 = \begin{bmatrix} \dot{x}_1 + r\dot{\gamma}_1 \cos\alpha_1 \\ \dot{y}_1 + r\dot{\gamma}_1 \sin\alpha_1 \\ \dot{z}_1 \\ \dot{\beta}_1 \\ \dot{\alpha}_1 \sin\beta_1 \\ \dot{\gamma}_1 + \dot{\alpha}_1 \cos\beta_1 \end{bmatrix} = \begin{bmatrix} w_{1x} \\ w_{1y} \\ w_{1z} \\ w_{1\beta} \\ w_{1\alpha} \\ w_{1\gamma} \end{bmatrix}. \qquad (2.59)$$

The fact that slippage is ignored leads to the following conditions:

$$w_{1x} = 0, \quad w_{1y} = 0, \quad w_{1z} = 0. \qquad (2.60)$$

The linear velocity of the frame mass centre in the $H\xi_2\eta_2\zeta_2$ frame, where its length is expressed in the $Hx_2y_2z_2$ frame and the wheel is represented by the $Hx_1y_1z_1$ frame, is as follows:

$$\boldsymbol{v}_{21} = \boldsymbol{R}_{\alpha 2}^T \boldsymbol{R}_{\gamma 1}\boldsymbol{v}_1 + \boldsymbol{\Omega}_2 \boldsymbol{R}_{\alpha 1}^T \boldsymbol{r}_{21}, \qquad (2.61)$$

where:

$$\boldsymbol{\Omega}_2 = \boldsymbol{R}_2^T \dot{\boldsymbol{R}}_2, \quad \boldsymbol{R}_2 = \boldsymbol{R}_{\alpha 1}\boldsymbol{R}_{\beta 1}\boldsymbol{R}_{\alpha 2}. \qquad (2.62)$$

Comparably to the frame mass centre velocity, the velocity of the saddle itself (it is placed at the end of the frame) is given by:

$$\boldsymbol{v}_2 = \boldsymbol{R}_{\alpha 2}^T \boldsymbol{R}_{\gamma 1}\boldsymbol{v}_1 + \boldsymbol{\Omega}_2 \boldsymbol{R}_{\alpha 1}^T \boldsymbol{r}_2, \qquad (2.63)$$

The quasi-velocities defining the frame model velocities (Fig. 2.4) are assumed in the form:

$$\boldsymbol{w}_2 = \boldsymbol{R}_{\alpha 2}\boldsymbol{\omega}_2 = \begin{bmatrix} w_{1\alpha} \\ w_{1\beta} \\ w_{2\alpha} \end{bmatrix}, \qquad (2.64)$$

The velocity of the unicycle trunk was obtained on the basis of the given scheme as well.

It is the power of the unicyclist's legs that acts as the drive, or still better, the propulsion of the mechanism under consideration. The motion of his/her extremities defines the position, conduct or movement of the whole system. However, in order to calculate the velocity and position of legs, the assumption was made, according to which the unicycle wheel and frame defined the position of unicyclist's extremities (this chapter). Such a hypothesis was formulated so that the relations between the limbs and the rest of the system could be determined. In the given case, the procedure is very much alike, despite the fact that no quasi-velocities occur. The quasi-velocity of each link in lower limbs is a resultant of the unicycle wheel and frame quasi-velocities.

Fig. 2.3 Wheel coordinates (green) and quasi-velocities (pink)

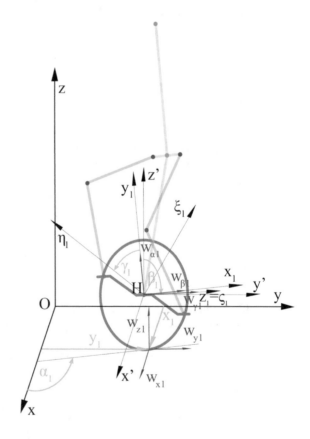

Fig. 2.4 Frame coordinates (green) and quasi-velocities (pink)

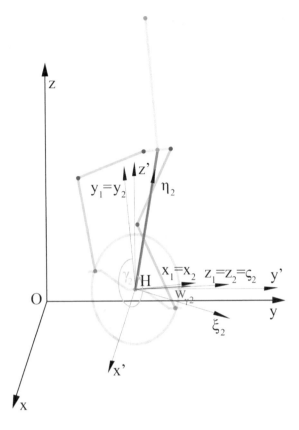

The quasi-velocities which define the unicycle-unicyclist model velocities (Figs. 2.3, 2.4, 2.5, 2.6 and 2.7) are presented in the following form:

$$
w = \begin{bmatrix} w_{1x} \\ w_{1y} \\ w_{1z} \\ w_{1\alpha} \\ w_{1\beta} \\ w_{1\gamma} \\ w_{2\alpha} \\ w_{3\alpha} \\ w_{3\beta} \end{bmatrix} = \begin{bmatrix} 1 & 0 & 0 & 0 & 0 & r\cos\alpha_1 & 0 & 0 & 0 \\ 0 & 1 & 0 & 0 & 0 & r\sin\alpha_1 & 0 & 0 & 0 \\ 0 & 0 & 1 & 0 & 0 & 0 & 0 & 0 & 0 \\ 0 & 0 & 0 & 0 & 1 & 0 & 0 & 0 & 0 \\ 0 & 0 & 0 & \sin\beta_1 & 0 & 0 & 0 & 0 & 0 \\ 0 & 0 & 0 & \cos\beta_1 & 0 & 1 & 0 & 0 & 0 \\ 0 & 0 & 0 & \cos\beta_1 & 0 & 0 & 1 & 0 & 0 \\ 0 & 0 & 0 & 0 & 0 & 0 & 0 & 0 & 1 \\ 0 & 0 & 0 & 0 & 0 & 0 & 0 & 1 & 0 \end{bmatrix} \begin{bmatrix} \dot{x}_1 \\ \dot{y}_1 \\ \dot{z}_1 \\ \dot{\alpha}_1 \\ \dot{\beta}_1 \\ \dot{\gamma}_1 \\ \dot{\alpha}_2 \\ \dot{\alpha}_3 \\ \dot{\beta}_3 \end{bmatrix} = A\dot{q} \quad (2.65)
$$

Noting that the gravitational forces are invariable, they can be included in the equations as a sum of the appropriate potential energy for each of the particular links

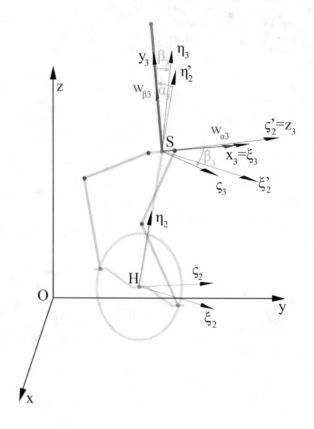

Fig. 2.5 Body coordinates (green) and quasi-velocities (pink)

of the unicycle-unicyclist system [16, 22, 23]. The potential energy given in the $Oxyz$ frame of the system can be expressed by the formula:

$$V = \sum_{i=1}^{7} m_i g y_i, \qquad i = (1, \ldots, 7).$$

(2.66)

An impact of the damping qualities exerted by synovial fluids in the appropriate joints [16, 22, 23] is given by the formula:

$$f_{i+1} = d_{i+1} (w_{i+1} - w_i) \qquad i = (1, 2, 3, 4, 5, 6, 7),$$

(2.67)

and included in the motion equations (Eq. (2.68)) as the external forces, similarly to the control force in Eq. (2.74).

Fig. 2.6 Right tibia
coordinates (green, left is
analogous)

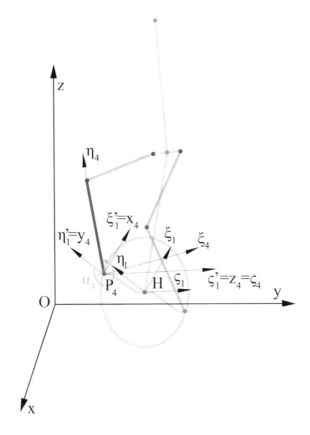

2.5   Equations of the Model Dynamics

The unicycle-unicyclist model is truly sophisticated. The system is characterised by 6 second-order partial differential equations. The presence of 6 equations instead of 9 (the number of general coordinates, this chapter) results from omitting the slippage factor in consideration. In the light of the previous derivations mentioned in Sect. 2.4, the motion equations are presented in the most transparent, uncomplicated and brief form. In such a case, the character of equations is clearly visible (Figs. 2.8, 2.9, 2.10, 2.11, 2.12, 2.13, 2.14 and 2.15).

Fig. 2.7 Right thigh coordinates (green, left is analogous)

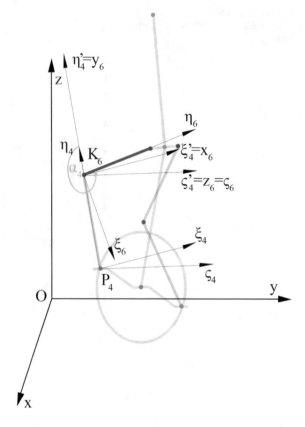

Fig. 2.8 Position of the right leg defined by the angles γ_1 and α_2

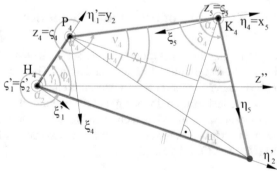

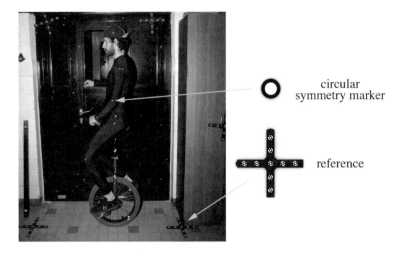

Fig. 2.9 3D body scan

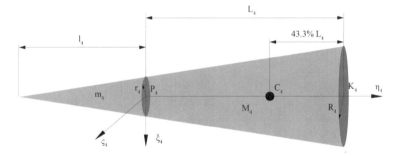

Fig. 2.10 Shape and mass center of the unicyclist's thigh

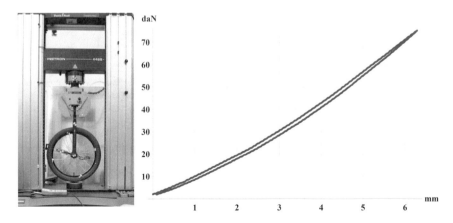

Fig. 2.11 Measurement of the tyre stiffness coefficient k_t

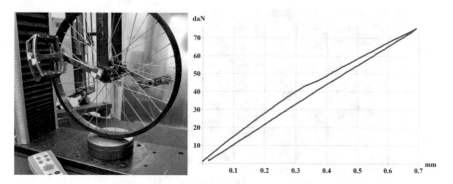

Fig. 2.12 Measurement of the rim stiffness coefficient k_r

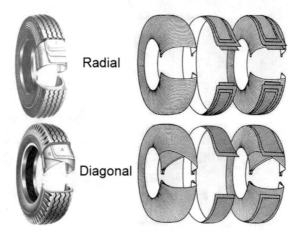

Fig. 2.13 Comparison of radial and diagonal tyres on the example of car tyres [15]

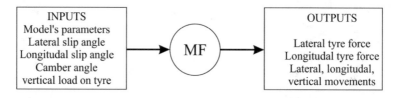

Fig. 2.14 Driving wheel parameters and a scheme of the Magic Formula

Fig. 2.15 Controlled system

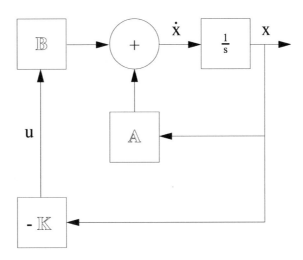

Let us introduce the model of the unicycle-unicyclist motion equations:

$$
\begin{cases}
\dfrac{d}{dt}\left(\dfrac{\partial T^*}{\partial w_{1\alpha}}\right) - \dfrac{\partial T^*}{\partial \beta_1}\csc(\beta_1) + \left(\dfrac{\partial T^*}{\partial w_{2\alpha}} + \dfrac{\partial T^*}{\partial w_{1\gamma}} - \dfrac{\partial T^*}{\partial w_{1\beta}}\cot(\beta_1)\right) w_\beta = \\[2mm]
= f_{1\beta} - \dfrac{\partial V}{\partial \beta_1} \\[3mm]
\dfrac{d}{dt}\left(\dfrac{T^*}{\partial w_{1\beta}}\right) + \left(\dfrac{\partial T^*}{\partial w_{1x}}\sin(\alpha_1) - \dfrac{\partial T^*}{\partial w_{1y}}\cos(\alpha_1)\right) R_1 \csc(\beta_1)\, w_{1\gamma} + \\[2mm]
+ \left(\dfrac{\partial T^*}{\partial w_{1\beta}}\cot(\beta_1) - \dfrac{\partial T^*}{\partial w_{1\gamma}} - \dfrac{\partial T^*}{\partial w_{2\alpha}}\right) w_{1\alpha} - \dfrac{\partial T^*}{\partial \alpha_1} = \\[2mm]
= \left(f_{1\alpha} - \dfrac{\partial V}{\partial \alpha_1}\right)\csc(\beta_1) + \left(\left(\left(f_{1x} - \dfrac{\partial V}{\partial x_1}\right) + \left(f_{1y} - \dfrac{\partial V}{\partial y_1}\right)\right) R_1 \sin(\alpha_1) + \\[2mm]
- \left(f_{2\alpha} - \dfrac{\partial V}{\partial \alpha_2}\right) - \left(f_{1\gamma} - \dfrac{\partial V}{\partial \gamma_1}\right)\right)\cot(\beta_1) \\[3mm]
\dfrac{d}{dt}\left(\dfrac{\partial T^*}{\partial w_{1\gamma}}\right) + \dfrac{\partial T^*}{\partial \beta_1}\cot(\beta_1) - \dfrac{\partial T^*}{\partial \gamma_1} + \\[2mm]
+ \left(\dfrac{\partial T^*}{\partial w_{1y}}\cos(\alpha_1) - \dfrac{\partial T^*}{\partial w_{1x}}\sin(\alpha_1)\right) R_1 \csc(\beta_1)\, w_{1\beta} = \\[2mm]
= f_\gamma - \dfrac{\partial V}{\partial \gamma_1} - \left(\left(f_{1x} - \dfrac{\partial V}{\partial x_1}\right)\cos(\alpha_1) + \left(f_{1y} - \dfrac{\partial V}{\partial y_1}\right)\sin(\alpha_1)\right) R_1 \\[3mm]
\dfrac{d}{dt}\left(\dfrac{\partial T^*}{\partial w_{2\alpha}}\right) + \dfrac{\partial T^*}{\partial \beta_1}\cot(\beta_1) - \dfrac{\partial T^*}{\partial \alpha_2} = f_{2\alpha} - \dfrac{\partial V}{\partial \alpha_2} \\[3mm]
\dfrac{d}{dt}\left(\dfrac{\partial T^*}{\partial w_{3\alpha}}\right) - \dfrac{\partial T^*}{\partial \beta_3} = f_{3\beta} - \dfrac{\partial V}{\partial \beta_3} \\[3mm]
\dfrac{d}{dt}\left(\dfrac{\partial T^*}{\partial w_{3\beta}}\right) - \dfrac{\partial T^*}{\partial \alpha_3} = f_{3\alpha} - \dfrac{\partial V}{\partial \alpha_3}
\end{cases}
\tag{2.68}
$$

2.6 Control System

In practice, it is the cyclist that bears responsibility for maintaining the vertical position of the unicycle-unicyclist system by controlling the driving wheel (control of speed and turns).

In the model, the desired optimal control force is generated by an explanation of the standard Linear–Quadratic Regulator (*LQR*) problem. Its main idea is to formulate a feedback control law to minimize the quadratic cost function being related to the matrices $\mathbb{Q}$ and $\mathbb{R}$. What turns out is that regardless of the $\mathbb{Q}$ and $\mathbb{R}$ values, the cost function reaches the unique minimum. The two parameters mentioned can be used as the design ones to penalize the state variables and the control signals. The larger these values are, the more the signals are penalized. The trade-off occurs between $\mathbb{Q}$ and $\mathbb{R}$. The strategies of choosing appropriate $\mathbb{Q}$ and $\mathbb{R}$ [24–35] are as follows:

- expensive control—considerable values of $\mathbb{R}$ (less energy involved),
- cheap control—limited values of $\mathbb{R}$ (more energy),
- changes in states—considerable values of $\mathbb{Q}$,
- the least changes in states—limited values of $\mathbb{Q}$.

The feedback control law:

$$u = -\mathbb{K}x. \tag{2.69}$$

The quadratic cost function for a continuous time system with the infinite horizon:

$$\mathbb{J}(u) = \int_0^\infty \underbrace{\left(x^T \mathbb{Q} x + u^T \mathbb{R} u\right)}_{\mathbb{L}(x,u)} dt. \tag{2.70}$$

is subject to the system dynamics:

$$\dot{x} = \mathbb{A}x + \mathbb{B}u. \tag{2.71}$$

The strategies for choosing the values of $\mathbb{Q}$ and $\mathbb{R}$ are as follows:

1. Vary ρ to get an expected response

$$\mathbb{Q} = \mathbb{I}, \quad \mathbb{R} = \rho\mathbb{I}, \quad \Rightarrow \quad \mathbb{L} = ||x||^2 + \rho||u||^2, \tag{2.72}$$

2. Choose each q_i to the given equal effort for the same badness

$$\mathbb{Q} = \begin{bmatrix} q_1 & & \\ & \ddots & \\ & & q_j \end{bmatrix}, \tag{2.73}$$

acceptable error $1\,\text{cm} \Rightarrow q_j = \left(\frac{1}{100}\right)^2 \quad q_j x_j^2 = 1$ when $x_j = 1\,\text{m}$
Similarly with $\mathbb{R}$, vary ρ to adjust the input/state balance.
3. The sheerly commonplace approach—by trial and error.

Control Systems, a specialised area of *Wolfram Mathematica*™, was applied to generate the control of the investigated system. At the outset, a model of the system behaviour near the equilibrium point was constructed. This can be done automatically as well, with the *StateSpace Model* function which depends on motion equations. The StateSpace Model resembles an expanded matrix. Every column describes an effect of slight derivations from equilibrium of each of the state variables. The expanded second column similarly describes an immediate effect of the control force on the state variables. The next step consists in examining the values for feedback gains by an implementation of the *LQRegulatorGains* function which is based on the $\mathbb{Q}$ and $\mathbb{R}$ parameters [36, 37]. With the gains, generated via *LQRegulatorGains*, the control force applied to the unicycle wheel can be presented in the form:

$$F_c = g_1\alpha_1 + g_2\beta_1 + g_3\gamma_1 + g_4\alpha_2 + g_5\alpha_3 + g_6\beta_3 +$$
$$+ g_7\dot{\alpha}_1 + g_8\dot{\beta}_1 + g_9\dot{\gamma}_1 + g_{10}\dot{\alpha}_2 + g_{11}\dot{\alpha}_3 + g_{12}\dot{\beta}_3, \qquad (2.74)$$

where each angle and angular velocity has its own gain. The given force was applied to Eq. (2.68). The behaviour of the model with the control added is demonstrated in Chap. 3.

References

1. Zenkov, D., Bloch, A., Leonard, N. E., & Marsden, J. E. (2000). Matching and stabilization of the unicycle with ridler. In *Conference on Decision and Control, IFAC Proceedings Volumes* (Vol. 33, pp. 177–178).
2. Naveh, Y., Bar-Yoseph, P., & Halevi, Y. (1999). Nonlinear modeling and control of a unicycle. *Dynamics and Control, 9*, 279–296.
3. Vos, D., & von Flotow, A. (1990). Dynamics and nonlinear adaptive control of an autonomous unicycle: Theory and experiment. In *Proceedings of the 33rd Conference on Decision and Control*.
4. Zhiyu, S., & Daliang, L. (2010). Balancing control of a unicycle riding. In *29th Chinese control conference*. IEEE.
5. Tsai, C., Chan, C., Shih, S., & Lin, S. (2008). Adaptive nonlinear control using RBFNN for an electric unicycle. In *IEEE international conference on system, man and cybernetics*. Singapore: SMC.
6. Micaelli, A., & Samson, C. (1993). *Trajectory tracking for unicycle-type and two-steering-wheels mobile robots* (Research Report RR-2097), INRIA. Inria-00074575.
7. Retrieved March, 2015, from https://www.impactunicycles.com/.
8. Retrieved November, 2015, from http://www.gom.com/3d-software/download.htm.
9. Clauser, C. E., McConville, J. T., & Young, J. W. (1971). Weight, volume, and center of mass of segments of the human body. *Journal of Occupational and Environmental Medicine, 13*, 270–280.
10. Arus, E. (2012). *Biomechanics of human motion*. Boca Raton: CRC Press.

11. Unsworth, A. (1991). Tribology of human and artificial joints. *Proc. Inst. Mech. Eng H.*, *205*, 163–172.
12. Retrieved March, 2016, from http://www.instron.us/en-us/products/testing-systems/universal-testing-systems/electromechanical.
13. Retrieved March, 2016, from http://www.maxxis.com/catalog/tire-246-radiale-22c.
14. Pacejka, H. (2006). *Tyre and vehicle dynamics*. Elsevier, Oxford: Butterworth-Heinemann.
15. Retrieved March, 2016, from https://www.tyreleader.co.uk/tyres-advices/number-tyre-plies.
16. Tongue, B. H. (2002). *Principles of vibration*. Oxford: Oxford University Press.
17. Talamucci, F. (2015). An algebraic procedure for reducing the Boltzmann hamel equations in nonholonomic systems. *Advances in Theoretical and Applied Mechanics*, *8*, 7–26.
18. Gutowski, R. (1992). Analytical mechanics, in: Foundations of Mechanics. In H. Zorski (Ed.), *PWN/Elsevier, Warsaw* (pp. 1–119). Warszawa: PWN.
19. Nejmark, J. I., & Fufajew, N. A. (1972). Dynamics of nonholonomic systems. *American Mathematical Society Providence*.
20. Blajer, W. (1995). Projective formulation of Lagrange's and Bolzmann-Hamel equations for multibody systems. *ZAMM*, *75*, S15–S108.
21. Maruskin, J. M., & Bloch, A. M. (2011). The Boltzmann's Hamel equations for the optimal control of mechanical systems with nonholonomic constraints. *Int. J. Robust. Nonlinear Control*, *21*, 373–386.
22. Grabski, J., & Strzałko, J. (1997). Automatic generation of Bolzmann-Hamel coefficients for mechanical problems. *Mechanics and Mechanical Engineering International Journal*, *1*, 59–76.
23. Grabski, J., Mianowski, B., & Strzako, J. *Wybrane zagadnienia mechaniki*. ISBN 83-87198-64-1.
24. Retrieved October, 2016, from https://www.researchgate.net/post/how.to.determine.the.values.of.the.control.matrices.Q.and.R.for.the.LQR.strategy.when.numerically.simulating.the.semi-active.TLCD.
25. Murray, R. (2006). *Lecture 2–LQR control*. California: Institute of Technology. www.cds.calltech.edu/murray/courses/cds110/wi06/lqr.pdf.
26. Sinclair, A., Hurtado, J., & Junkins, J. (2006). Linear feedback control using quasi velocities. *Journal of Guidance, Control and Dynamics*, *29*, 1309–1314.
27. Lee, T., Song, K., & Lee, C. (2001). Tracking control of unicycle-modeled mobile robots using a saturation feedback controller. *IEEE Transactions on Control System Technology*, *9*, 305–318.
28. Kim, B., & Tsiotras, P. (2002). Controllers for unicycle-type wheeled robots: Theoretical results and experimental validation. *IEEE Transactions on Robotics and Automation*, *18*, 294–307.
29. Aicardi, M., Casalino, G., Balestrino, A., & Bicchi, A. (1994). Closed loop smooth steering of unicycle-like vehicles. In *Proceedings of the 33rd Conference on Decision and Control*.
30. Mellors, M., & College, P. (2005). *Robotic unicycle: Mechanics and control* (Technical Milestone Report). www.roboticunicycle.info.
31. Kinoshita, M., Yoshida, K., Sugimoto, Y., Ohsaki, H., Yoshida, H., Iwase, M., & Hatakeyama, S. (2008). Support control to promote skill of riding a unicycle. In *IEEE international conference on system, man and cybernetics*. Singapore: SMC.
32. Ohsaki, H., Iwase, M., Sadahiro, T., & Hatakeyama, S. (2009). A consideration of human-unicycle model for unicycle operation analysis based on moment balancing point. In *IEEE international conference on system, man and cybernetics*. San Antonio: SMC.
33. Shao, Z., & Liu, D. (2010). Balancing control of a unicycle riding. In *29-th Chinese control conference CCC 2010*. IEEE.
34. Ulyanov, S., Watanabe, S., Ulyanov, V., Yamafuji, K., Litvintseva, L., & Rizzotto, G. (1998). Soft computing for the intelligent robust control of a robotic unicycle with a new physical measure for mechanical controllability. *Soft Computing*, *2*, 13–88.

35. Sheng, Z., & Yamafuji, K. (1995). Study on the stability and motion control of a unicycle: Part I: Dynamics of a human riding a unicycle and its modelling by link mechanics. JSME International Journal. *Ser. C, Dynamics, control, robotics and manufacturing, 38*, 249–259.
36. Moylan A. Stabilized inverted pendulum. Retrieved Jan, 2017, from http://blog.wolfram.com/2011/01/19/stabilized-inverted-pendulum/.
37. Moylan A. Stabilized n-Link pendulum. Retrieved Jan, 2017, from http://blog.wolfram.com/2011/03/01/stabilized-n-link-pendulum/.

Chapter 3
Numerical and Experimental Validation of the Model

Validation is a documented procedure aimed at confirming that all the processes, equipment, materials, procedures, activities and systems actually lead to the planned results. First, the model developed in Chap. 2 was validated in the numerical simulations and next, on the basis of the experimental data obtained during a real unicycle ride. Due to the complexity of the system, before the 3-D validation was used, the 2-D one had been performed in order to verify the direction of investigations and simulations.

3.1 Numerical Simulations

On the basis of the model developed in Chap. 2, the *Wolfram Mathematica*™code given in Appendix 2 was implemented. The visualization of the numerical simulation for the complete unicycle–unicyclist model is depicted in Figs. 3.1, 3.2, 3.3, 3.4, 3.5, 3.6, 3.7 and 3.8. The main initial conditions related to the simulation comprised the constant velocity of the wheel as well as its vertical position.

The calculations were performed for the following set values of parameters: $R_1 = 0.25\,\text{m}$, $r_1 = 0.03\,\text{m}$, $M_1 = 3.8\,\text{kg}$, $r_2 = 0.005\,\text{m}$, $l_2 = 0.3\,\text{m}$, $L_2 = 0.65\,\text{m}$, $M_2 = 1.4\,\text{kg}$, $d_2 = 0.001$, $R_3 = 0.22\,\text{m}$, $r_3 = 0.07\,\text{m}$, $L_3 = 0.81\,\text{m}$, $M_3 = 43.5\,\text{kg}$, $d_3 = 0.003$, $R_4 = 0.056\,\text{m}$, $r_4 = 0.036\,\text{m}$, $L_4 = 0.421\,\text{m}$, $M_4 = 3.1\,\text{kg}$, $d_4 = 0.002$, $R_5 = 0.079\,\text{m}$, $r_5 = 0.052\,\text{m}$, $L_5 = 0.485\,\text{m}$, $M_5 = 7.2\,\text{kg}$, $d_5 = 0.003$, $R_6 = 0.056\,\text{m}$, $r_6 = 0.036\,\text{m}$, $L_6 = 0.421\,\text{m}$, $M_6 = 3.2\,\text{kg}$, $d_6 = 0.002$, $R_7 = 0.079\,\text{m}$, $r_7 = 0.052\,\text{m}$, $L_7 = 0.485\,\text{m}$, $M_7 = 7.2\,\text{kg}$, $d_7 = 0.003$, $\rho = 1050\,\text{kg/m}^3$, $g = 9.81\,\text{m/s}^2$, and the initial conditions: $x_1(0) = 3.2$, $y_1(0) = 0.02$, $z_1(0) = 0$, $\alpha_1(0) = 0$, $\beta_1(0) = \pi/100$, $\gamma_1(0) = 0$, $\alpha_2(0) = -\pi/33$, $\alpha_3(0) = -\alpha_2(0)$, $\beta_3(0) = 0$, $t = 1.1\,\text{s}$, $\dot{\alpha}_1(0) = 0$, $\dot{\beta}_1(0) = 0$, $\dot{\gamma}_1(0) = 9\,\text{rad/s}$, $\dot{\alpha}_2(0) = 0$, $\dot{\alpha}_3(0) = 0$, $\dot{\beta}_3(0) = 0$ (we used the parameters of a QU-AX Luxus 20 unicycle).

© The Author(s) 2019
M. Niełaczny et al., *Dynamics of the Unicycle*,
SpringerBriefs in Applied Sciences and Technology,
https://doi.org/10.1007/978-3-319-95384-7_3

Fig. 3.1 Time dependence of the wheel contact point with and without the tyre model in the dimensions: **a** x, **b** y and **c** z

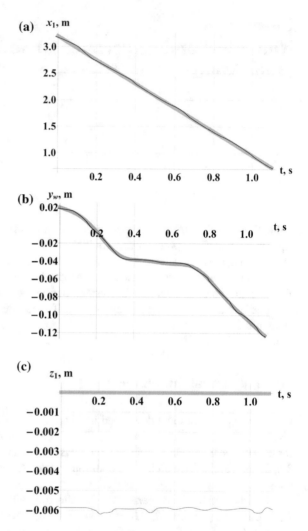

The thick yellow lines represent the displacement in each of the dimensions of the contact point, situated between the driving wheel and the ground, according to the assumption that slippage is omitted, Eq. (2.60). The delicate black lines exemplify how the position changes in relation to the particular contact point dimension, for the model equipped with a pneumatic tyre. It seems perfectly vivid that the impact of a pneumatic tyre in the motion under consideration is insignificant.

According to Fig. 3.2a, the whole system, visually represented by the wheel alone, performs slight but constant left or right turns. What is more, Fig. 3.2b implies that the wheel is subjected to meagre left or right deflections from the vertical position as well. Figure 3.2c shows a depiction of the wheel rolling forwards.

Fig. 3.2 Time dependence of the wheel angles of rotation: **a** α_1, **b** β_1 and **c** γ_1

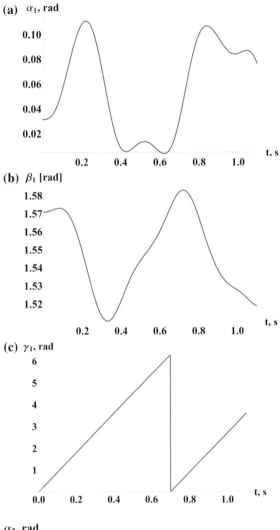

Fig. 3.3 Time dependence of the frame angle of rotation α_2

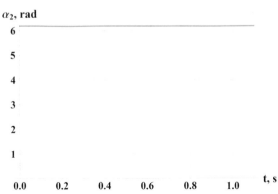

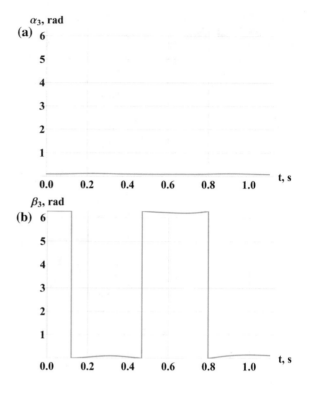

Fig. 3.4 Time dependence of the trunk angle of rotation: **a** α_3 and **b** β_3

It is clearly shown in Fig. 3.3 that the frame is deflected backwards by approximately $\pi/33$ rad from the vertical and maintains such a position. It is characteristic of unicycles because of the contribution of lower extremities, which serve as the counterweight. The purpose of all the presented factors is to keep the mass centre above the fulcrum.

Additionally, the trunk is slightly deflected as well, but forwards rather than backwards and to a lesser extent in comparison to the frame. It leans towards the side determined by the direction of travel, as can be seen in (Fig. 3.4a). Apart from that, the trunk swings from left to right (Fig. 3.4b), contrarily to β_1.

Figures 3.5a and b and Figs. 3.6a and b depict rotation angles of the lower extremities' links in the plane parallel to the wheel plane. The right leg is evenly shifted in phase in comparison to the left one by the value of $\frac{\pi}{2}$. The reason for this phenomenon is the phase shift of the cranks by the value of $\frac{\pi}{2}$.

The same results as in Fig. 3.1, 3.2, 3.3, 3.4, 3.5 and 3.6 are shown in Fig. 3.7, as 3D trajectories in the frame $Oxyz$ of characteristic points such as:

- fulcrum—a curve oscillating around the straight line on the base,
- hub—a curve oscillating around the straight line in space,
- pedals—the trajectories resemble curtate cycloids shifted in phase due to the fact that the cranks are phase-shifted as well, by the value of $\dfrac{\pi}{2}$,

Fig. 3.5 Time dependence of the angle of rotation of: **a** right thigh—α_4 and **b** left thigh—α_6

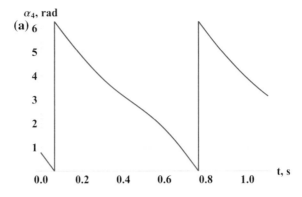

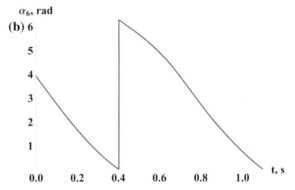

- knees—the trajectories resemble periodic functions shifted in phase due to the fact that the cranks are phase-shifted by the value of $\dfrac{\pi}{2}$,
- saddle—a curve oscillating around the straight line in space,
- head—a curve oscillating around the straight line in space.

The above results perfectly correspond to the unicycle riding description (Sect. 1.1) and are validated in Sect. 3.2.2.

The correctness of the simulation results is presented in Fig. 3.8, which depicts changes in values of the quasi-velocities w_1, w_2 and w_3 in the course of the simulation process.

The equation error of constraints is presented by means of the values of quasi-velocities in the contact point which ought to be equal to 0, bearing in mind that slippage is ignored in motion for an exact solution. The maximal values of the simulation error are settled at the level of 10^{-7} m/s.

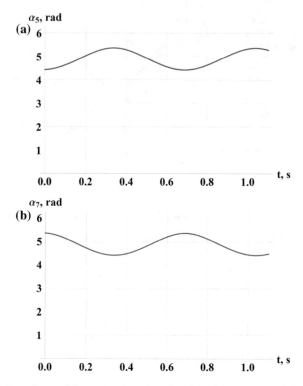

Fig. 3.6 Time dependence of the angle of rotation of: **a** right tibia—α_5 and **b** left tibia—α_7

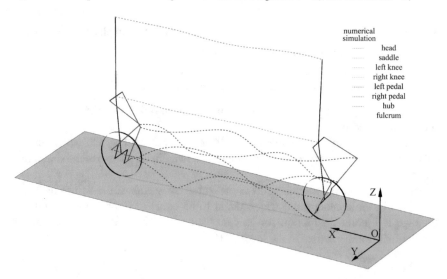

Fig. 3.7 Simulation of the unicycle–unicyclist model motion

Fig. 3.8 Time dependence of the quasi-velocity of the contact point in the dimensions: **a** x, **b** y and **c** z

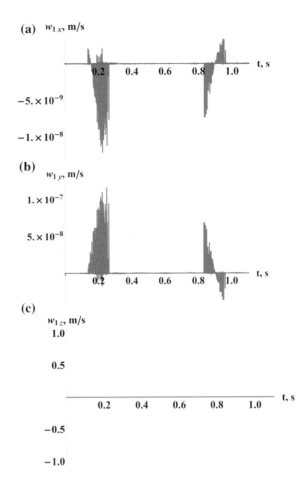

3.2 Experimental Validation of the Model

3.2.1 Experimental Validation

3.2.1.1 2-D Motion Capture

The *TEMA* Motion software is designed for advanced motion analysis. In order to track objects, *TEMA*, which is based on digital image sequences, carries out movement analyses automatically and presents the results in predefined formats such as tables and graphs. They depict quantifiable values, for instance position, speed or acceleration versus time. The operator is free to choose between a considerable number of subpixel tracking algorithms and follow/record an unlimited number of points [1] (Fig. 3.9).

Fig. 3.9 Location of quadrant symmetry markers on the unicycle

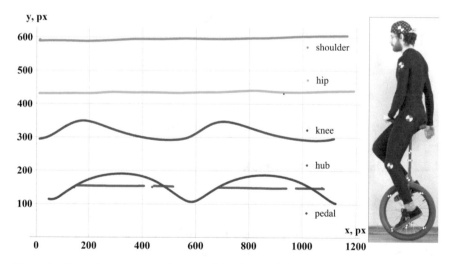

Fig. 3.10 2-D motion capture—trajectories of the characteristic points

Therefore, to capture the motion of an actual model, a high-speed camera—*Vision Research Phantom VR0908* was used. The movie was recorded in the 1000 fps mode. A single attempt duration amounts to two seconds. The *TEMA* software was used to process the movies. Figure 3.10 depicts parametric plots of the positions of model characteristic plots, as regards the hub, pedal, knee, hip and shoulder.

After that experiment, the direction of investigations was confirmed and the behaviour of "quick-and-dirty" characteristic points was obtained.

3.2.2 3-D Motion Capture

OptiTrack Motive: Track is a software platform designed for following the objects in 6-DoF with exact precision for various tracking applications. Not only does Motive enable one to calibrate and configure the system, but also to capture and process the 3-D data. *Motive* obtains 3-D coordinates via 2-D images of markers. Through using the 3-D coordinates described by the tracked markers, *OptiTrack Motive: Track* is able to acquire six-degree-of-freedom data for multiple rigid bodies and skeletons (scaffolding) by an application of the *Motive:Body* plug. Moreover, it enables tracking complex movements in 3-D [2, 3] (Fig. 3.11).

The $3D$ validation was introduced to certify the proposed 3-D model (Fig. 2.1). In order to capture the motion of a given model in three dimensions, *OptiTrack Flex 13*, a system consisting of six high-speed cameras, was brought into operation. The movies were recorded in a 120 fps mode. To specify the dimensions of the motion capture area, it was 3 m high, 3.5 m long and 3 m wide, which appeared to be the maximal range in accordance with the manufacturer's recommendations. Two seconds is an average duration of a single attempt. Apart from the reflective ones, thirty-seven markers were fastened to the unicyclist's body. It was done due to the requirements of the *OptiTrack Motive: Body* software so that it could detect a human. Eight markers were attached to the unicycle, four of which were fastened to the frame and four to the wheel. The markers' location on the unicycle was optimized so as not to be covered by the cyclist's extremities, the frame, cranks or spokes when moving. To be the most precise, the marker attached to the rim at one side was covered in a flash once per every full rotation cycle. The *OptiTrack Motive: Body* software was used for processing the movies. Figure 3.12 depicts a comparison of the parametric plots between the experiments and the simulations of characteristic points connected to the unicycle–unicyclist model (Fig. 2.1):

- pedal,
- knee,
- hip,
- shoulder.

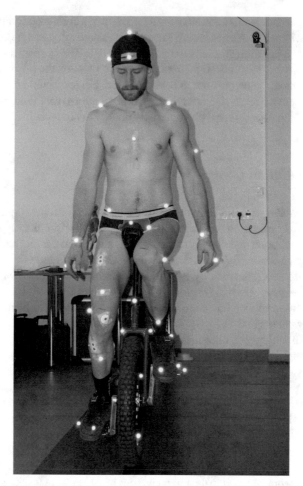

Fig. 3.11 Location of reflective markers

Figure 3.12 corresponds directly to Fig. 3.7 supplemented with the 3-D motion
capture experimental trajectories. Comparably to Fig. 3.7, the dashed lines represent
the numerical solution, whereas the solid transparent lines show the experiment
results. The numerical results reflect the behaviour of real objects directly.

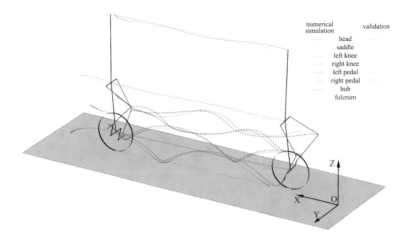

Fig. 3.12 3-D model validation

References

1. Image Systems–Tema. Retrieved Nov, 2015, from http://www.imagesystems.se/index.php/tema/.
2. OptiTrack. Retrieved Oct, 2016, from http://www.optitrack.com/.
3. OptiTrack Wiki Documentation. Retrieved Oct, 2016, from http://wiki.optitrack.com/index.php?title=MotiveDocumentation.

Chapter 4
Concluding Remarks

The first author was delighted to investigate his point of interest from a scientific perspective and to convince everyone that passion and science can go hand in hand.

The matrix notation of the Boltzmann–Hamel equations eliminates the drawbacks associated with their classic formulation. An application of these equations facilitates the automation of generation of motion equations. Derivation of the matrix which transforms the generalized velocities into quasi-velocities, the kinetic energy and the vector of generalized forces is an ample movement aimed at obtaining the equations of motion in the form of quasi-coordinates and quasi-velocities. The foregoing procedure is commonplace and can be used for solving many different problems.

The described model depicts the conduct of a unicyclist when riding a unicycle. It was validated and compatibility between the simulations and the experiment was obtained. The model facilitates simulations of the unicyclist's movements. After all, the impact of the tyre, having been taken into consideration at the outset, turned out to be insignificant, which was confirmed on the basis of the experience and the simulations. As a result of that phenomenon, a further simplification, due to which the system of the second-order differential equations was reduced from 9 to 6, could be introduced. It results from the fact that slippage was omitted in the course of consideration. The concept of the unicycle–unicyclist model control system is outlined in the form of the drive wheel guidance system.

The model of human lower extremities based on the biomechanical rules highlights the most significant difficulty related to unicycling, i.e., an increase in the wheel imbalance caused by the function of cranks and pedals. It is the motion of legs that hinders learning to ride so dramatically, although it does not attract attention of the majority of authors [1–7]. It is probably caused by the ignorance of scientists taking up the topic, as they are not active unicyclists.

© The Author(s) 2019
M. Niełaczny et al., *Dynamics of the Unicycle*,
SpringerBriefs in Applied Sciences and Technology,
https://doi.org/10.1007/978-3-319-95384-7_4

Due to the suggested control system, the model's behaviour corresponds to the real object, which makes it more animated. It was confirmed via the simulations and experience. Having in mind the absence of an interaction with the unicyclist's trunk, its motion was treated as a noise. Maintaining the unicycle's frame angular position as well as sustaining the model mass centre above the fulcrum in the unstable equilibrium were brought into sharp focus.

The numerical model properly mirrors the real object at the present stage of research and successfully meets the main targets of the book. The differences between the proposed model and the real object may result from the fact that:

- the model is both idealization and simplification of the reality;
- the maximal range of the motion capture working area was matched to the manufacturer's recommendations. This could be the reason for skipping the markers, as the cameras could not initially detect the unicyclist crossing the examination area. Therefore, he had to begin his ride within the working area in order to get clear courses;
- the unicyclist experienced slight deflections when reaching the point of inertia. He wobbles for a while before maintaining the upright position and keeping the straight track;
- it was assumed that the lower extremities moved in the plane which was parallel to one of the driving wheel. In fact, the knees might slightly swing from side to side during the ride in order to maintain the balance.

Future goals and paths of development will be aimed at:

- creating an application destined for mobile phones which could be at every unicyclist's disposal. The user would be able to adjust the initial conditions, for instance the unicycle's or unicyclist's parameters as well as to load a track profile, in order to verify the simulation of motion. In such a case, the applied tyre model with configurable pressure would be justified;
- bike fitting, which is becoming more and more popular. It is nothing but an optimization of the bicycle parameters and position of the rider to suit his/her preferences and abilities, in order to achieve the best results through a little effort. Nowadays, it concerns only regular bikes. The application mentioned in the previous point could exert a great influence on the unicycle fitting;
- the given model that will serve as the proper basis for development of a self-balancing unicycle, which could be appropriate for each user;
- creating a follow-up system which will balance cranks, pedals and cyclist's legs and which would make learning to ride easier and accessible for the disabled;
- such a system that will enable the unicyclist to reach higher speed values, which could turn a unicycle into a common means of transport convenient in traffic;
- extending the unicyclist's model by adding feet and arms to make it more human-like.

References

1. Zenkov, D., & Bloch, A. (1999). Stabilization of the unicycle with rider. In *Conference on Decision and Control, IFAC Proceedings Volumes* (Vol. 33, pp. 177–178).
2. Naveh, Y., Bar-Yoseph, P., & Halevi, Y. (1999). Nonlinear modeling and control of a unicycle. *Dynamics and Control, 9*, 279–296.
3. Vos D., & von Flotow A. (1990). Dynamics and nonlinear adaptive control of an autonomous unicycle: Theory and experiment. *Decision and Control*.
4. Zhiyu, S., & Daliang, L. (2010). Balancing control of a unicycle riding. In *IEEE of the 29th Chinese Control Conference*.
5. Tsai C., Chan C., Shih S., & Lin, S. (2008). Adaptive nonlinear control using RBFNN for an electric unicycle. In *IEEE International Conference on Systems, Man and Cybernetics, SMC Singapore 2008*.
6. Mellors, M., & College, P. (2005). Robotic unicycle: Mechanics and control. Technical Milestone Report (www.roboticunicycle.info).
7. Kinoshita, M., Yoshida, K., Sugimoto, Y., Ohsaki, H., Yoshida, H., Iwase, M., & Hatakeyama, S. (2008). Support control to promote skill of riding a unicycle. In *IEEE International Conference on Systems, Man and Cybernetics, SMC Singapore 2008*.

Appendix A
Euler Angles

In order to determine the position of the rigid body, the frame which is to move together with the body is bound to it. The rigid body position is evaluated unambiguously when the position of the point ∇ and the direction of the axes of the frame are known with reference to the fixed $0xyz$ frame. This phenomenon allows one to determine the location of each particular point of the given body. Several ways to describe the orientation of the coordinate system associated with the body can be enumerated. Euler angles as well as their modifications such as Bryant (Cardan) angles seem to be most commonly used [1–4].

The initial spatial orientation of the body is defined by the position of the frame $\nabla x'y'z'$, being parallel to the main frame $0xyz$. Subsequently, three rotations are arranged, and, as a result, the rigid body is transformed into the position defined by the frame associated with the body—$\nabla \xi_i \eta_i \zeta_i$ (Fig. A.1).

A description of the orientation of the coordinate system associated with the rigid body with Euler angles [2] is as follows:

$$
\begin{array}{ccc}
rotation & axis & angle \\
1 & z' & \alpha_i \\
2 & x_{0i} & \beta_i \\
3 & z_i & \gamma_i
\end{array}
\qquad (A.1)
$$

Equations A.2–A.4 shows the transformation matrices:

$$
\boldsymbol{R}_{\alpha_i} = \begin{bmatrix} \cos\alpha_i & -\sin\alpha_i & 0 \\ \sin\alpha_i & \cos\alpha_i & 0 \\ 0 & 0 & 1 \end{bmatrix}, \qquad (A.2)
$$

$$
\boldsymbol{R}_{\beta_i} = \begin{bmatrix} 1 & 0 & 0 \\ 0 & \cos\beta_i & -\sin\beta_i \\ 0 & \sin\beta_i & \cos\beta_i \end{bmatrix}, \qquad (A.3)
$$

© The Author(s) 2019
M. Niełaczny et al., *Dynamics of the Unicycle*,
SpringerBriefs in Applied Sciences and Technology,
https://doi.org/10.1007/978-3-319-95384-7

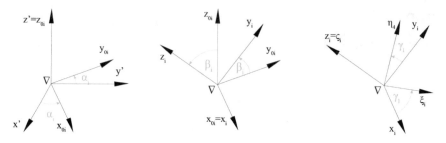

Fig. A.1 Initial position of Euler angles and subsequent rotations around the axis: z', x_{0i}, z_i [4]

$$R_{\gamma_i} = \begin{bmatrix} \cos\gamma_i & -\sin\gamma_i & 0 \\ \sin\gamma_i & \cos\gamma_i & 0 \\ 0 & 0 & 1 \end{bmatrix}. \qquad (A.4)$$

In this case, the inverse matrices $R_{\alpha_i}^{-1}$, $R_{\beta_i}^{-1}$, $R_{\gamma_i}^{-1}$ are equal to the transpose matrices $R_{\alpha_i}^{T}$, $R_{\beta_i}^{T}$, $R_{\gamma_i}^{T}$:

$$R_i^{-1} \equiv R_i^T. \qquad (A.5)$$

Equation A.6 shows how to represent a coordinate within the frame $\nabla x_{0i} y_{0i} z_{0i}$ with a coordinate within the frame $\nabla \xi_i \eta_i \zeta_i$:

$$x_{0i} = R_{\alpha_i} R_{\beta_i} R_{\gamma_i} \xi_i, \qquad (A.6)$$

and a similar conversion from $\nabla x_{0i} y_{0i} z_{0i}$ to $\nabla \xi_i \eta_i \zeta_i$:

$$\xi_i = R_{\gamma_i}^T R_{\beta_i}^T R_{\alpha_i}^T x_{0i}. \qquad (A.7)$$

Euler angles are commonly used to describe the rigid body motion.

Appendix B
Wolfram Mathematica™ Code for Dynamics of the Unicycle-Unicyclist System

```
(* ROTATION MATRICES *)
Rα[n_]:={{Cos[x[n][t]], −Sin[x[n][t]], 0}, {Sin[x[n][t]], Cos[x[n][t]], 0}, {0, 0, 1}};
Rα[n]//MatrixForm;
Rβ[n_]:={{1, 0, 0}, {0, Cos[x[n + 1][t]], −Sin[x[n + 1][t]]},
   {0, Sin[x[n + 1][t]], Cos[x[n + 1][t]]}};
Rβ[n]//MatrixForm;
Rγ[n_]:={{Cos[x[n + 2][t]], −Sin[x[n + 2][t]], 0},
   {Sin[x[n + 2][t]], Cos[x[n + 2][t]], 0}, {0, 0, 1}};
Rγ[n]//MatrixForm;
(* ROTATION MATRICES *)

(*COORDINATES*)
z[n][t] = z_n; y[n][t] = y_n; x[n][t] = x_n;
x[n][t] = α_n; x[n + 1][t] = β_n; x[n + 2][t] = γ_n;
Style[TableForm[{{1, 2, 3}, {z_{1n}, x_{2n}, z_{3n}}, {α[n][t], β[n][t], γ[n][t]},
   {MatrixForm[Rα[n]], MatrixForm[Rβ[n]], MatrixForm[Rγ[n]]}},
TableHeadings → {{"rotation", "axis", "angle", "matrix"}, None},
TableAlignments → {Center}], {Bold, Orange}]//TraditionalForm
(*COORDINATES*)
(*WHEEL*)
(* ROTATION MATRICES *)
Rα[4]; Rβ[4]; Rγ[4];
RHO1 = FullSimplify[Rα[4].Rβ[4]];
RHO2 = FullSimplify[Rα[4].Rβ[4].Rγ[4]];
ROH = FullSimplify[Transpose[Rγ[4].Transpose[Rβ[4].Transpose[Rα[4]]]]];
(*VECTORS*)
rOT = {{x[1][t]}, {x[2][t]}, {x[3][t]}};
rTH = {{0}, {r}, {0}};
rTH2 = {{0}, {r/2}, {0}};
```

© The Author(s) 2019
M. Niełaczny et al., *Dynamics of the Unicycle*,
SpringerBriefs in Applied Sciences and Technology,
https://doi.org/10.1007/978-3-319-95384-7

rOH = rOT + RHO1.rTH;
rOH2 = rOT + RHO1.rTH2;
(*VELOCITIES*)
vT = vTw = {{0}, {0}, {0}}; (* T as wheel point *)
ΩHH = Transpose[RHO2].D[RHO2, t];
(*inH$\xi_w \eta_w \zeta_w$*)
ωH = {{ΩHH[[3, 2]]}, {ΩHH[[1, 3]]}, {ΩHH[[2, 1]]}};
wH = Rγ[4].ωH;
vH = vT + Partition[Cross[Flatten[ωH], Transpose[Rγ[4]].Flatten[rTH]], 1];
(*inquasiveloinH$\xi_w \eta_w \zeta_w$*)
wHw = {{w[4][t]}, {w[5][t]}, {w[6][t]}};
ωHw = FullSimplify[Transpose[Rγ[4]].wHw];
vHw = FullSimplify[vT+
    Partition[Cross[Flatten[ωHw], Transpose[Rγ[4]].Flatten[rTH]], 1]];
(*WHEEL*)

(*HUB*)
(*VECTORS*)
rHR = $\left\{ \{0\}, \{r\}, \left\{ -\frac{h}{2} \right\} \right\}$;
rHL = $\left\{ \{0\}, \{r\}, \left\{ \frac{h}{2} \right\} \right\}$;
rOHR = rOT + RHO1.rHR;
rOHL = rOT + RHO1.rHL;
(* TO LEGS MOTION *)
rHR2 = $\left\{ \{0\}, \{r\}, \left\{ -\frac{h}{2} - \frac{p}{2} \right\} \right\}$;
rHL2 = $\left\{ \{0\}, \{r\}, \left\{ \frac{h}{2} + \frac{p}{2} \right\} \right\}$;
rOHR2 = rOT + RHO1.rHR2;
rOHL2 = rOT + RHO1.rHL2;
(*HUB*)

(*CRANKS*)
(*VECTORS*)
rHCR = $\left\{ \{0\}, \{c\}, \left\{ -\frac{h}{2} \right\} \right\}$;
rHCL = $\left\{ \{0\}, \{-c\}, \left\{ \frac{h}{2} \right\} \right\}$;
rHCL2 = $\left\{ \{0\}, \left\{ -\frac{c}{2} \right\}, \left\{ \frac{h}{2} \right\} \right\}$;
rOCR = rOH + RHO2.rHCR;
rOCL = rOH + RHO2.rHCL;
rOCL2 = rOH + RHO2.rHCL2;
(*CRANKS*)

(*PEDALS*)
(*VECTORS*)
rHPR = $\left\{ \{0\}, \{c\}, \left\{ -\frac{h}{2} - p \right\} \right\}$;
rHPL = $\left\{ \{0\}, \{-c\}, \left\{ \frac{h}{2} + p \right\} \right\}$;
rOPR = rOH + RHO2.rHPR;

```
rOPL = rOH + RHO2.rHPL;
(* TO LEGS MOTION *)
```
$$\text{rHPR2} = \left\{\{0\}, \{c\}, \left\{-\tfrac{h}{2} - \tfrac{p}{2}\right\}\right\};$$
$$\text{rHPL2} = \left\{\{0\}, \{-c\}, \left\{\tfrac{h}{2} + \tfrac{p}{2}\right\}\right\};$$
```
rOPR2 = rOH + RHO2.rHPR2;
rOPL2 = rOH + RHO2.rHPL2;
(*VELOCITIES*)
```
$(*\text{inH}\xi_w \eta_w \zeta_w*)$
```
(*vPR = Partition[Cross[Flatten[ωH], Transpose[Rγ[4]].(Flatten[rHPR2]
+Flatten[rTH])], 1]*)
vPR = vT+
    Partition[Cross[Flatten[ωH], Transpose[Rγ[4]].Flatten[rHPR2]], 1];
vPL = vT+
    Partition[Cross[Flatten[ωH], Transpose[Rγ[4]].Flatten[rHPL2]], 1];
```
$(*\text{inquasiveloinH}\xi_w \eta_w \zeta_w*)$
```
vPRw = vTw+
    Partition[Cross[Flatten[ωHw], Transpose[Rγ[4]].Flatten[rHPR2]], 1];
vPLw = vTw+
Partition[Cross[Flatten[ωHw], Transpose[Rγ[4]].Flatten[rHPL2]], 1];
(*PEDALS*)

(*FRAME*)
(* ROTATION MATRIX *)
Rγ[5];
RFO = RHO1.Rγ[5];
(*VECTORS*)
rHS = {{0}, {f}, {0}};
```
$$\text{rHF} = \left\{\{0\}, \left\{\tfrac{f}{2}\right\}, \{0\}\right\};$$
```
rOS = rOH + RFO.rHS;
rOF = rOH + RFO.rHF;
(*VELOCITIES*)
ΩFF = Transpose[RFO].D[RFO, t];
```
$(*\text{inF}\xi_f \eta_f \zeta_f*)$
```
ωF = {{ΩFF[[3, 2]]}, {ΩFF[[1, 3]]}, {ΩFF[[2, 1]]}};
wF = Rγ[5].ωF;
vF = Transpose[Rγ[5]].Rγ[4].vH+
    Partition[Cross[Flatten[ωF], Transpose[Rγ[5]].Flatten[rHF]], 1];
vS = Transpose[Rγ[5]].Rγ[4].vH+
    Partition[Cross[Flatten[ωF], Transpose[Rγ[5]].Flatten[rHS]], 1];
```
$(*\text{inquasiveloinH}\xi_f \eta_f \zeta_f*)$
```
wFw = {{w[4][t]}, {w[5][t]}, {w[7][t]}};
ωFw = FullSimplify[Transpose[Rγ[5]].wFw];
vFw = FullSimplify[Transpose[Rγ[5]].Rγ[4].vHw+
    Partition[Cross[Flatten[ωFw], Transpose[Rγ[5]].Flatten[rHF]], 1]];
```

```
vSw = FullSimplify[Transpose[Rγ[5]].Rγ[4].vHw+
  Partition[Cross[Flatten[ωFw], Transpose[Rγ[5]].Flatten[rHS]], 1]];
(*FRAME*)

(*HIPS*)
(*VECTORS*)
rSR = {{0}, {f}, {-h/2 - p/2}};
rSL = {{0}, {f}, {h/2 + p/2}};
rOSR = rOH + RFO.rSR;
rOSL = rOH + RFO.rSL;
(*HIPS*)

(*BODY*)
(* ROTATION MATRIX *)
RBF = Rβ[7].Rγ[7];
RBO = RFO.RBF;
(*VECTORS*)
rSB = {{0}, {b}, {0}};
rSB2 = {{0}, {b/2}, {0}};
rOB = rOS + RBO.rSB;
rOB2 = rOS + RBO.rSB2;

(*VELOCITIES*)
ΩBB = Transpose[RBO].D[RBO, t];
(*inBξ_bη_bζ_b*)
ωB = {{ΩBB[[3, 2]]}, {ΩBB[[1, 3]]}, {ΩBB[[2, 1]]}};
wB = Rγ[7].ωB;
vB = Transpose[Rγ[7]].Transpose[Rβ[7]].Rγ[5].vS+
  Partition[Cross[Flatten[ωB], Transpose[Rγ [7]].Flatten[rSB]], 1];
(*inquasiveloinBξ_bη_bζ_b*)
wBw = {{Cos[x[7][t]]w[4][t] + Sin[x[7][t]]w[5][t] + w[8][t]},
  {(-Cos[x[8][t]]Sin[x[7][t]])w[4][t]
    + (Cos[x[7][t]]Cos[x[8][t]])w[5][t]+
  Sin[x[8][t]]w[7][t]}, {Sin[x[7][t]]Sin[x[8][t]]w[4][t]
    - Cos[x[7][t]]Sin[x[8][t]]w[5][t]
    + Cos[x[8][t]]w[7][t] + w[9][t]}};
ωBw = FullSimplify[Transpose[Rγ [7]].wBw];
vBw = FullSimplify[Transpose[Rγ[7]].Transpose[Rβ[7]].Rγ[5].vSw
  + Partition[Cross[Flatten[ω Bw], Transpose[Rγ[7]].Flatten[rSB]], 1]];
(*BODY*)
```

```
(*LEGS*)
(*RIGHT*)


;
(*ANGLES OF ROTATION DEPENDENT ON φR*)
φR[t] = x[6][t] − x[7][t];
PRSR = Sqrt[(c * Sin[φR[t]])^2 + (f − (c * Cos[φR[t]]))^2];
μR[t] = ArcTan[f − (c * Cos[φR[t]]), c * Sin[φR[t]]];
χR[t] = ArcCos[(−th^2 + PRSR^2 + ti^2)/(2 * ti * PRSR)];
δR[t] = ArcCos[(−PRSR^2 + th^2 + ti^2)/(2 * ti * th)];
νR[t] = χR[t] − μR[t];
ψR[t] = 2π − (φR[t] − νR[t]);
θR[t] = 2π − (π − δR[t]);
(*ANGLES OF ROTATION DEPENDENT ON φR*)


(*TIBIA*)
(* ROTATION MATRIX *)
Rα[10];
(*x[11][t] = ψR[t]; *)
RTIRO = RHO2.Rα[10];
(*VECTORS*)
rPTI = {{0}, {ti/2}, {0}} ;
rPK = {{0}, {ti}, {0}};
rOTIR = rOPR2 + RTIRO.rPTI;
rOKR = rOPR2 + RTIRO.rPK;
(*VELOCITIES*)
ΩTIRTIR = Transpose[RTIRO].D[RTIRO, t];
(*inTIrξ_tir η_tir ζ_tir *)
ωTIR = {{ΩTIRTIR[[3, 2]]}, {ΩTIRTIR[[1, 3]]}, {ΩTIRTIR[[2, 1]]}};
wTIR = Rα[10].ωTIR;
vTIR = Transpose[Rα[10]].vPR+
   Partition[Cross[Flatten[ωTIR], Transpose[Rα[10]].Flatten[rPTI]], 1];
vKR = Transpose[Rα[10]].vPR+
   Partition[Cross[Flatten[ωTIR], Transpose[Rα[10]].Flatten[rPK]], 1];
```

$$AA =$$

$$\left(c\left(cti\sqrt{c^2 + f^2 - 2cf\,\text{Cos}[x[6][t] - x[7][t]]} \right.\right.$$

$$\sqrt{\left(-\left(c^4 + (f - th - ti)(f + th - ti)(f - th + ti)(f + th + ti) + \right.\right.}$$

$$c^2\left(4f^2 - 2\left(th^2 + ti^2\right)\right) +$$

$$2cf\left(2\left(-c^2 - f^2 + th^2 + ti^2\right)\text{Cos}[x[6][t] - x[7][t]] + \right.$$

$$\left.\left. cf\,\text{Cos}[2(x[6][t] - x[7][t])])\right) / \right.$$

$$\left.\left.\left(ti^2\left(c^2 + f^2 - 2cf\,\text{Cos}[x[6][t] - x[7][t]]\right)\right)\right) - \right.$$

$$f \, \text{ti} \, \text{Cos}[x[6][t] - x[7][t]] \sqrt{c^2 + f^2 - 2cf\text{Cos}[x[6][t] - x[7][t]]}$$
$$\sqrt{\left(-\left(c^4 + (f - \text{th} - \text{ti})(f + \text{th} - \text{ti})(f - \text{th} + \text{ti})(f + \text{th} + \text{ti}) + \right.\right.}$$
$$c^2 \left(4f^2 - 2\left(\text{th}^2 + \text{ti}^2\right)\right) +$$
$$2cf \left(2\left(-c^2 - f^2 + \text{th}^2 + \text{ti}^2\right)\text{Cos}[x[6][t] - x[7][t]] + \right.$$
$$\left.cf\text{Cos}[2(x[6][t] - x[7][t])]\right)) /$$
$$\left(\text{ti}^2 \left(c^2 + f^2 - 2cf\text{Cos}[x[6][t] - x[7][t]]\right)\right)) -$$
$$f \left(c^2 + f^2 + \text{th}^2 - \text{ti}^2 - 2cf\text{Cos}[x[6][t] - x[7][t]]\right) \text{Sin}[x[6][t] - x[7][t]]) /$$
$$\left(\text{ti} \left(c^2 + f^2 - 2cf\text{Cos}[x[6][t] - x[7][t]]\right)^{3/2}\right.$$
$$\sqrt{\left(-\left(c^4 + (f - \text{th} - \text{ti})(f + \text{th} - \text{ti})(f - \text{th} + \text{ti})(f + \text{th} + \text{ti}) + \right.\right.}$$
$$c^2 \left(4f^2 - 2\left(\text{th}^2 + \text{ti}^2\right)\right) +$$
$$2cf \left(2\left(-c^2 - f^2 + \text{th}^2 + \text{ti}^2\right)\text{Cos}[x[6][t] - x[7][t]] + cf\text{Cos}[2(x[6][t] - x[7][t])]\right)) /$$
$$\left(\text{ti}^2 \left(c^2 + f^2 - 2cf\text{Cos}[x[6][t] - x[7][t]]\right)\right)));$$

BB =

$$\left(f\right.$$
$$\left(f \, \text{ti} \sqrt{c^2 + f^2 - 2cf\text{Cos}[x[6][t] - x[7][t]]}\right.$$
$$\sqrt{\left(-\left(c^4 + (f - \text{th} - \text{ti})(f + \text{th} - \text{ti})(f - \text{th} + \text{ti})(f + \text{th} + \text{ti}) + \right.\right.}$$
$$c^2 \left(4f^2 - 2\left(\text{th}^2 + \text{ti}^2\right)\right) +$$
$$2cf \left(2\left(-c^2 - f^2 + \text{th}^2 + \text{ti}^2\right)\text{Cos}[x[6][t] - x[7][t]] + cf\text{Cos}[2(x[6][t] - x[7][t])]\right)) /$$
$$\left(\text{ti}^2 \left(c^2 + f^2 - 2cf\text{Cos}[x[6][t] - x[7][t]]\right)\right)) -$$
$$c \, \text{ti} \, \text{Cos}[x[6][t] - x[7][t]] \sqrt{c^2 + f^2 - 2cf\text{Cos}[x[6][t] - x[7][t]]}$$
$$\sqrt{\left(-\left(c^4 + (f - \text{th} - \text{ti})(f + \text{th} - \text{ti})(f - \text{th} + \text{ti})(f + \text{th} + \text{ti}) + \right.\right.}$$
$$c^2 \left(4f^2 - 2\left(\text{th}^2 + \text{ti}^2\right)\right) +$$
$$2cf \left(2\left(-c^2 - f^2 + \text{th}^2 + \text{ti}^2\right)\text{Cos}[x[6][t] - x[7][t]] + cf\text{Cos}[2(x[6][t] - x[7][t])]\right)) /$$
$$\left(\text{ti}^2 \left(c^2 + f^2 - 2cf\text{Cos}[x[6][t] - x[7][t]]\right)\right)) +$$
$$c \left(c^2 + f^2 + \text{th}^2 - \text{ti}^2 - 2cf\text{Cos}[x[6][t] - x[7][t]]\right) \text{Sin}[x[6][t] - x[7][t]]) /$$
$$\left(\text{ti} \left(c^2 + f^2 - 2cf\text{Cos}[x[6][t] - x[7][t]]\right)^{3/2}\right.$$
$$\sqrt{\left(-\left(c^4 + (f - \text{th} - \text{ti})(f + \text{th} - \text{ti})(f - \text{th} + \text{ti})(f + \text{th} + \text{ti}) + \right.\right.}$$
$$c^2 \left(4f^2 - 2\left(\text{th}^2 + \text{ti}^2\right)\right) +$$
$$2cf \left(2\left(-c^2 - f^2 + \text{th}^2 + \text{ti}^2\right)\text{Cos}[x[6][t] - x[7][t]] + cf\text{Cos}[2(x[6][t] - x[7][t])]\right)) /$$
$$\left(\text{ti}^2 \left(c^2 + f^2 - 2cf\text{Cos}[x[6][t] - x[7][t]]\right)\right)));$$

(*in quasivelo in TIr$\xi_{\text{tir}}\eta_{\text{tir}}\zeta_{\text{tir}}$*)
    wTIRw = {{Sin[x[6][t]]w[5][t] + Cos[x[6][t]]w[4][t]},
    {Cos[x[6][t]]w[5][t] - Sin[x[6][t]]w[4][t]},
    {w[10][t](*Cos[x[5][t]]x[4]'[t] + AAx[6]'[t] + BBx[7]'[t]*)}} /.{w[10][t] → 0};
ωTIRw = Transpose[Rα[10]].wTIRw/.{w[10][t] → 0}; ;
vTIRw = Transpose[Rα[10]].vPRw+
    Partition[Cross[Flatten[ωTIRw], Transpose[Rα[10]].Flatten[rPTI]], 1]/.{w[10][t]
        → 0}; ;
vKRw = Transpose[Rα[10]].vPRw+
    Partition[Cross[Flatten[ωTIRw], Transpose[Rα[10]].Flatten[rPK]], 1]/.{w[10][t]
        → 0}; ;
(*TIBIA*)
(*LEGS*)

```
(*V(*WHEEL*)
Vw = mwgrOH[[3]];
(*FRAME*)
Vf = mfgrOF[[3]];
(*BODY*)
Vb = mbgrOB2[[3]];
(* TIBIA R*)
Vtir = mtigrOTIR[[3]];
(*SYSTEM*)
V = Vw + Vf + Vb + Vtir;
(*V*)


(*MASS*)
M[n_]:={{m[n], 0, 0}, {0, m[n], 0}, {0, 0, m[n]}};
m[1] = mr; (*rim*)
m[2] = mh; (*hub*)
m[3] = mc; (*crank*)
m[4] = mp; (*pedal*)
mw = mr + mh + 2mc + 2mp;
m[5] = mf; (*frame*)
m[6] = mb; (*body*)
m[7] = mti; (*tibia*)
MR = M[1];
MH = M[2];
MC = M[3];
MP = M[4];
MW = MR + MH + 2MC + 2MP;
MF = M[5];
MB = M[6];
MTI = M[7];
(*MASS*)

(*MOMENT OF INERTIA *)
MOI[n_]:=
    {{Iξ[n], Iξη[n], Iξζ[n]}, {Iξη[n], Iη[n], Iηζ[n]}, {Iξζ[n], Iηζ[n], Iζ[n]}};
(*MOMENT OF INERTIA *)

(*WHEEL*)
(*WHEELVELOCITYINHξwηwζw*)
ωHwr = [ωHw];
vHwr = Transpose[vHw];
(*WHEEL VELOCITY IN Hξwηwζw*)
(*MOMENT OF INERTIA IN IN Hξwηwζw*)
```

```
(*RIM*)
```
$I\xi[1] = \frac{mrr^2}{2};$
$I\xi\eta[1] = 0;$
$I\xi\zeta[1] = 0;$
$I\xi\eta[1] = 0;$
$I\eta[1] = \frac{mrr^2}{2};$
$I\eta\zeta[1] = 0;$
$I\xi\zeta[1] = 0;$
$I\eta\zeta[1] = 0;$
$I\zeta[1] = mrr^2;$
```
MOIR = MOI[1];
(*HUB*)
```
$I\xi[2] = \frac{mhh^2}{12};$
$I\xi\eta[2] = 0; I\xi\zeta[2] = 0;$
$I\xi\eta[2] = 0;$
$I\eta[2] = \frac{mhh^2}{12};$
$I\eta\zeta[2] = 0;$
$I\xi\zeta[2] = 0;$
$I\eta\zeta[2] = 0;$
$I\zeta[2] = 0;$
```
MOIH = MOI[2];

(*CRANKS*)
```
$I\xi[3] = \left(\frac{mcc^2}{12} + mc\left(\left(\frac{h}{2}\right)^2 + \left(\frac{c}{2}\right)^2\right)\right) + \left(\frac{mc(-c)^2}{12} + mc\left(\left(-\frac{h}{2}\right)^2 + \left(-\frac{c}{2}\right)^2\right)\right);$
$I\xi\eta[3] = 0;$
$I\xi\zeta[3] = 0;$
$I\xi\eta[3] = 0;$
$I\eta[3] = \left(0 + mc\left(\frac{h}{2}\right)^2\right) + \left(0 + mc\left(-\frac{h}{2}\right)^2\right);$
$I\eta\zeta[3] = \left(mc\frac{c}{2}\frac{h}{2}\right) + \left(mc\left(-\frac{c}{2}\right)\left(-\frac{h}{2}\right)\right);$
$I\xi\zeta[3] = 0;$
$I\eta\zeta[3] = \left(mc\frac{c}{2}\frac{h}{2}\right) + \left(mc\left(-\frac{c}{2}\right)\left(-\frac{h}{2}\right)\right);$
$I\zeta[3] = \left(\frac{mcc^2}{12} + mc\left(\frac{c}{2}\right)^2\right) + \left(\frac{mc(-c)^2}{12} + mc\left(-\frac{c}{2}\right)^2\right);$
```
MOIC = MOI[3];
(*PEDALS*)
```
$I\xi[4] = \left(\frac{mpp^2}{12} + mp\left(\left(\frac{h}{2} + \frac{p}{2}\right)^2 + c^2\right)\right) + \left(\frac{mp(-p)^2}{12} + mp\left(\left(-\frac{h}{2} - \frac{p}{2}\right)^2 + (-c)^2\right)\right);$
$I\xi\eta[4] = 0;$
$I\xi\zeta[4] = 0;$
$I\xi\eta[4] = 0; I\eta[4] = \left(\frac{mpp^2}{12} + mp\left(\frac{h}{2} + \frac{p}{2}\right)^2\right) + \left(\frac{mp(-p)^2}{3} + mp\left(-\frac{h}{2} - \frac{p}{2}\right)^2\right);$
$I\eta\zeta[4] = \left(mpc\left(\frac{h}{2} + \frac{p}{2}\right)\right) + \left(mp(-c)\left(-\frac{h}{2} - \frac{p}{2}\right)\right);$
$I\xi\zeta[4] = 0;$
$I\eta\zeta[4] = \left(mpc\left(\frac{h}{2} + \frac{p}{2}\right)\right) + \left(mp(-c)\left(-\frac{h}{2} - \frac{p}{2}\right)\right);$
$I\zeta[4] = \left(0 + mpc^2\right) + \left(0 + mpc^2\right);$
```
```

```
MOIP = MOI[4];
(*WHEEL*)
MOIW = MOIR + MOIH + MOIC + MOIP;
```
$(*$MOMENT OF INERTIA IN IN H$\xi_w \eta_w \zeta_w*)$

$(*$WHEEL $T*)$
TW = FullSimplify $\left[\frac{1}{2}(\text{vHwr.MW.vHw}) + \frac{1}{2}(\omega\text{Hwr.MOIW.}\omega\text{Hw})\right]$;
$(*$WHEEL $T*)$

$(*$FRAME$*)$
$(*$FRAME VELOCITY IN F$\xi_f \eta_f \zeta_f*)$
```
ωFwr = Transpose[ωFw];
vFwr = Transpose[vFw];
```
$(*$FRAME VELOCITY IN F$\xi_f \eta_f \zeta_f*)$

$(*$MOMENT OF INERTIA INF$\xi_f \eta_f \zeta_f*)$
I$\xi[5] = \frac{\text{mf}(f)^2}{12}$;
I$\xi\eta[5] = 0$;
I$\xi\zeta[5] = 0$;
I$\xi\eta[5] = 0$;
I$\eta[5] = 0$;
I$\eta\zeta[5] = 0$;
I$\xi\zeta[5] = 0$;
I$\eta\zeta[5] = 0$;
I$\zeta[5] = \frac{\text{mf}(f)^2}{12}$;
```
MOIF = MOI[5];
```
$(*$MOMENT OF INERTIA IN F$\xi_f \eta_f \zeta_f*)$
$(*$ FRAME $T*)$
TF = FullSimplify $\left[\frac{1}{2}(\text{vFwr.MF.vFw}) + \frac{1}{2}(\omega\text{Fwr.MOIF.}\omega\text{Fw})\right]/.\{x[10][t] \to 0\}$;
$(*$FRAME $T*)$
$(*$FRAME $*)$

$(*$BODY$*)$
$(*$BODY VELOCITY IN B$\xi_b \eta_b \zeta_b*)$
```
ωBwr = Transpose[ωBw];
vBwr = Transpose[vBw];
```
$(*$BODY VELOCITY IN B$\xi_b \eta_b \zeta_b*)$
$(*$MOMENT OF INERTIA IN B$\xi_b \eta_b \zeta_b*****)$
I$\xi[6] = \frac{\text{mb}(b)^2}{12}$; I$\xi\eta[6] = 0$; I$\xi\zeta[6] = 0$;
I$\xi\eta[6] = 0$; I$\eta[6] = 0$; I$\eta\zeta[6] = 0$;
I$\xi\zeta[6] = 0$; I$\eta\zeta[6] = 0$; I$\zeta[6] = \frac{\text{mb}(b)^2}{12}$;
```
MOIB = MOI[6];
```
$(*$MOMENT OF INERTIA IN B$\xi_b \eta_b \zeta_b*)$

(*BODY T*)
TB $= \frac{1}{2}$(vBwr.MB.vBw) $+ \frac{1}{2}$(ωBwr.MOIB.ωBw)/.$\{x[10][t] \to 0\}$;
(*BODY T*)
(*BODY*)

(*RIGHT TIBIA*)
(*RIGHT TIBIA VELOCITY IN Sr$\xi_{tir}\eta_{tir}\zeta_{tir}$*)
vTIRwr $=$ Transpose[vTIRw];
ωTIRwr $=$ Transpose[ωTIRw];
(*RIGHTTHIGHVELOCITYINTI$\xi_{ti}\eta_{ti}\zeta_{ti}$*)

(*MOMENT OF INERTIA IN TI$\xi_{ti}\eta_{ti}\zeta_{ti}$*)
I$\xi[8] = \frac{\text{mti(ti)}^2}{12}$; I$\xi\eta[8] = 0$; I$\xi\zeta[8] = 0$;
I$\xi\eta[8] = 0$; I$\eta[8] = 0$; I$\eta\zeta[8] = 0$;
I$\xi\zeta[8] = 0$; I$\eta\zeta[8] = 0$; I$\zeta[8] = \frac{\text{mti(ti)}^2}{12}$;
MOITI $=$ MOI[8];
(*MOMENT OF INERTIA IN TI$\xi_{ti}\eta_{ti}\zeta_{ti}$***)

(*RIGHT THIGH T*)
TTIR $= \frac{1}{2}$(vTIRwr.MTI.vTIRw) $+ \frac{1}{2}$(ωTIRwr.MOITI.ωTIRw)/.$\{x[10][t] \to 0\}$;
(*RIGHT THIGH T*)
(*RIGHT TIBIA *)

Print["DONE"]

rotation	1	2	3
axis	z_n	x_{2n}	z_{3n}
angle	$\alpha(n)(t)$	$\beta(n)(t)$	$\gamma(n)(t)$
matrix	$\begin{bmatrix} \cos(\alpha_n) & -\sin(\alpha_n) & 0 \\ \sin(\alpha_n) & \cos(\alpha_n) & 0 \\ 0 & 0 & 1 \end{bmatrix}$	$\begin{bmatrix} 1 & 0 & 0 \\ 0 & \cos(\beta_n) & -\sin(\beta_n) \\ 0 & \sin(\beta_n) & \cos(\beta_n) \end{bmatrix}$	$\begin{bmatrix} \cos(\gamma_n) & -\sin(\gamma_n) & 0 \\ \sin(\gamma_n) & \cos(\gamma_n) & 0 \\ 0 & 0 & 1 \end{bmatrix}$

DONE
TTIR

(* JG *)
ik=9;
$q =$ Table[$x[i][t]$, $\{i, 1, ik\}$];
$A = \{\{1, 0, 0, 0, 0, r\text{Cos}[x[4][t]], 0, 0, 0\}$,
$\{0, 1, 0, 0, 0, r\text{Sin}[x[4][t]], 0, 0, 0\}$,
$\{0, 0, 1, 0, 0, 0, 0, 0, 0\}$,
$\{0, 0, 0, 0, 1, 0, 0, 0, 0\}$,
$\{0, 0, 0, \text{Sin}[x[5][t]], 0, 0, 0, 0, 0\}$,
$\{0, 0, 0, \text{Cos}[x[5][t]], 0, 1, 0, 0, 0\}$,
$\{0, 0, 0, \text{Cos}[x[5][t]], 0, 0, 1, 0, 0\}$,
$\{0, 0, 0, 0, 0, 0, 0, 1, 0\}$,
$\{0, 0, 0, 0, 0, 0, 0, 0, 1\}$(*,

```
{0, 0, 0, 0, 0, 0, 0, 0, 0, 1}*)};
B = Simplify[Inverse[A]];
BT = Transpose[B];
MatrixForm[A]
MatrixForm[BT]
(*JG*)
```

(* TRANSFORMATION MATRICES *)

(* BOLTZMANN-HAMEL COEFFITIONS *)
```
DD[s_]:=Table[D[A[[s]], q[[i]]], {i, 1, ik}]
GG[s_]:=Simplify[BT.(DD[s] − Transpose[DD[s]]).B]
G = Table[GG[i], {i, 1, ik}];
MatrixForm[G];(* next matrices - row written *)
```
(* BOLTZMANN-HAMEL COEFFITIONS *)

(* QUASI-VELOCITIES VECTOR *)
```
ww = Table[w[i][t], {i, 1, ik}];
```
(* QUASI-VELOCITIES VECTOR *)

(* RELATIONS BETWEEN GENERALIZED COORDINATIONS
AND QUASI-VELOCITIES AND FORCE *)
```
Adq = A.D[q, t];
BTdVdq = BT.Table[D[V, x[i][t]], {i, 1, ik}];
```
(* RELATIONS BETWEEN GENERALIZED COORDINATIONS
AND QUASI- VELOCITIES AND FORCE *)

(*KINETIC ENERGY T *)
```
T = TW[[1, 1]] + TF[[1, 1]] + TB[[1, 1]] + TTIR[[1, 1]];
```
(*KINETIC ENERGY T *)

(*DERIVATIVES OF T *)
```
dTdw = Table[D[T, w[i][t]], {i, 1, ik}];
ddTdwdt = D[dTdw, t];
dTdq = Table[D[T, x[i][t]], {i, 1, ik}];
BTdTdq = dTdq.BT;
```
(*DERIVATIVES OF T *)

(*PARTS OF $B - H$ EQUATIONS*)
BOL = Transpose[Table[(Transpose[$G[[i]]$].ww), {i, 1, ik}]].dTdw;
(*PARTS OF $B - H$ EQUATIONS*)

(*$B - H$ EQUATIONS*)
BHE = Table[ddTdwdt[[i]] − BTdTdq[[i]] + BOL[[i]], {i, 1, ik}];
(*$B - H$ EQUATIONS*)
Print["DONE"]

$$\begin{pmatrix} 1\ 0\ 0 & 0 & 0 & r\mathrm{Cos}[x[4][t]]\ 0\ 0\ 0 \\ 0\ 1\ 0 & 0 & 0 & r\mathrm{Sin}[x[4][t]]\ 0\ 0\ 0 \\ 0\ 0\ 1 & 0 & 0 & 0\ 0\ 0 \\ 0\ 0\ 0 & 0 & 1 & 0 & 0\ 0\ 0 \\ 0\ 0\ 0\ \mathrm{Sin}[x[5][t]]\ 0 & 0 & 0\ 0\ 0 \\ 0\ 0\ 0\ \mathrm{Cos}[x[5][t]]\ 0 & 1 & 0\ 0\ 0 \\ 0\ 0\ 0\ \mathrm{Cos}[x[5][t]]\ 0 & 0 & 1\ 0\ 0 \\ 0\ 0\ 0 & 0 & 0 & 0 & 0\ 1\ 0 \\ 0\ 0\ 0 & 0 & 0 & 0 & 0\ 0\ 1 \end{pmatrix}$$

$$\begin{pmatrix} 1 & 0 & 0 & 0 & 0 & 0 & 0 & 0\ 0 \\ 0 & 1 & 0 & 0 & 0 & 0 & 0 & 0\ 0 \\ 0 & 0 & 1 & 0 & 0 & 0 & 0 & 0\ 0 \\ 0 & 0 & 0 & 0 & 1 & 0 & 0 & 0\ 0 \\ r\mathrm{Cos}[x[4][t]]\mathrm{Cot}[x[5][t]] & r\mathrm{Cot}[x[5][t]]\mathrm{Sin}[x[4][t]] & 0 & \mathrm{Csc}[x[5][t]] & 0 & -\mathrm{Cot}[x[5][t]] & -\mathrm{Cot}[x[5][t]]\ 0\ 0 \\ -r\mathrm{Cos}[x[4][t]] & -r\mathrm{Sin}[x[4][t]] & 0 & 0 & 0 & 1 & 0 & 0\ 0 \\ 0 & 0 & 0 & 0 & 0 & 0 & 1 & 0\ 0 \\ 0 & 0 & 0 & 0 & 0 & 0 & 0 & 1\ 0 \\ 0 & 0 & 0 & 0 & 0 & 0 & 0 & 0\ 1 \end{pmatrix}$$

DONE
(* DATA & INITIAL CONDITIONS *)
(*IMAGE*)
fontsize = 14;
imagesize = 500;
(*RIM*)
r = 0.3;
mr = 1.5;
(*HUB*)
mh = 0.1;
h = 0.12;
(*CRANK*)
mc = 0.3;
c = 0.137;
(*PEDAL*)
mp = 0.1;
p = 0.1;
(*WHEEL*)
mw = mr + mh + 2mc + 2mp;

(*FRAME*)
mf = 0.7;
$f = 0.65$;
(*BODY*)
mb = 0.000040; mb = 0.000040 * 100; (*JG *)
$b = 0.4$;
(*TIBIA*)
mti = 0.0000001; mti = 0.0000001 * 100; (*JG *)
ti = 0.4;
(*THIGH*)
th = 0.4;
(*GRAVITY*)
$g = 9.81$;
(*TIME*)
t0 = 0;
t1 = .2 * 5; t1 = .2 * 10;
maxstepsize = 10^{-3}; maxstepsize = 10^{-4};
(*INITIALCONDITIONS − POSITION*)
x0[1] = 0.;
x0[2] = 0.;
x0[3] = 0.;
x0[4] = 0.;
x0[5] = $\frac{\pi}{2}$;
x0[6] = 0.;
x0[7] = 0.;
x0[8] = 0.;
x0[9] = 0.;

(*INITIAL CONDITIONS − VELOCITY*)
dx0[1] = 0;
dx0[2] = 0;
dx0[3] = 0;
dx0[4] = 0.;
dx0[5] = 0.;
dx0[6] = 3.;
dx0[7] = $\frac{-r\,w0[6]}{\frac{L}{2}}$; dx0[7] = $\frac{-r(dx0[6]+Cos[x0[5]]dx0[4])}{\frac{L}{2}}$; (*JG*)
dx0[8] = 0;
dx0[9] = $\frac{-r\,w0[6]}{f+\frac{b}{2}}$; dx0[9] = $\frac{-r(dx0[6]+Cos[x0[5]]dx0[4])}{f+\frac{b}{2}}$; (*JG*)

(*INITIAL CONDITIONS − QUASIVELOCITY*)
w0[1] = 0;
w0[2] = 0;
w0[3] = 0;
w0[4] = dx0[5];

w0[5] = Sin[x0[5]]dx0[4];
w0[6] = dx0[6] + Cos[x0[5]]dx0[4];
w0[7] = dx0[7] + Cos[x0[5]]dx0[4];
w0[8] = dx0[8];
w0[9] = dx0[9];
(*w0[10] = dx0[10]; *)
(*INITIAL CONDITIONS*)

(*TABLE OF CONTENTS*)
Style [TableForm [{{g, "gravity", g, "m/s^2"} ,
{r, "wheel radius", r, m},
{h, "hub length", h, m},
{c, "crank length", c, m},
{p, "pedal length", p, m},
{f, "frame length", f, m},
{"mw", "wheel mass", mw, "kg"},
{"mh", "hub mass", mh, "kg"},
{"mc", "crank mass", mc, "kg"},
{"mp", "pedal mass", mp, "kg"},
{"mf", "frame mass", mf, "kg"},
{"mb", "frame mass", mb, "kg"},
{"x0[1]", " x_w ", x0[1], m} ,
{"x0[2]", " y_w ", x0[2], m} ,
{"x0[3]", " z_w ", x0[3], m} ,
{"x0[4]", "initial position α_w ", x0[4]180/Pi, "^{o}" } ,
{"x0[5]", " β_w ", x0[5]180/Pi, "^{o}" } ,
{"x0[6]", " γ_w ", x0[6]180/Pi, "^{o}" } ,
{"x0[7]", " α_f ", x0[7]180/Pi, "^{o}" } ,
{"x0[8]", " α_b ", x0[8]180/Pi, "^{o}" } ,
{"x0[9]", " β_b ", x0[9]180/Pi, "^{o}" } ,
{"dx0[4]", "initial velocity $\dot{\alpha}_w$ ", dx0[4], "rad"} ,
{"dx0[5]", " $\dot{\beta}_w$ ", dx0[5], "rad"} ,
{"dx0[6]", " $\dot{\gamma}_w$ ", dx0[6], "rad"} ,
{"dx0[7]", " $\dot{\alpha}_f$ ", dx0[7], "rad"} ,
{"dx0[7]", " $\dot{\alpha}_b$ ", dx0[8], "rad"} ,
{"dx0[5]", " $\dot{\beta}_b$ ", dx0[9], "rad"} ,
{"-", "linear velocity", $\frac{dx0[6]3600r}{1000}$, "rad"} ,
{"dt", "max step size", maxstepsize, s},
{"t0", "start time", t0, s},
{"t1", "stop time", t1, s},
{"Δt","motion time",t1–t0,s}},
TableHeadings → {None, {"symbol", "description", "value", "unit"}},
    TableAlignments → Center], {Bold, Orange}]

```
(*TABLE OF CONTENTS*)


(*MOTION EQUATIONS*)

ME = Flatten[{
Table[BHE[[i]]== − BTdVdq[[i, 1]], {i, {4, 5, 6, 7, 8, 9(*,10*)}}],
Table[w[i][t]==Adq[[i]], {i, {4, 5, 6, 7, 8, 9(*,10*)}}],
Adq[[1]]==0, Adq[[2]]==0, Adq[[3]]==0,
Table[w[i][0]==w0[i], {i, {4, 5, 6, 7, 8, 9}}],
Table[x[i][0]==x0[i], {i, 1, ik}]}]/.{x[10][t] → 0, w[10][t] → 0};
(*SUBSTITUTION*)

(*SUBSTITUTION*)
(*MOTION EQUATIONS*)


(*QUASI − VELOCITIES EQUATION*)
QVE = Flatten[{Table[w[i][t], {i, 1, ik}], Table[q[[i]], {i, 1, ik}]}];
(*QUASI − VELOCITIES EQUATION*)


(*TIME*)
TIME = {t, t0, t1};
(*TIME*)

Print[YYYY]
(*SOLVE*)
(*SOL = NDSolve[ME, QVE, TIME, Method →
   Automatic(*"Ex0licitRungeKutta"*),
   MaxStepSize → maxstepsize, MaxSteps → Infinity]; *)
SOL = NDSolve[ME, QVE, TIME, Method →
   {"EquationSimplification"->"Residual"},
   MaxStepSize → maxstepsize, MaxSteps → Infinity];
(*SOLVE*)
Print[DONE]


(*QUASI-VELOCITIES PLOTS*)
(*W1*)
PLOTW1 = Plot[Evaluate[r D[x[6][t]/.SOL, t]Cos[x[4][t]/.SOL]+
   D[x[1][t]/.SOL, t]], {t, t0, t1},
   GridLines->Automatic, AxesLabel->{"t [s]", "w1"},
PlotLabel->"quasi velocity w1 = r γ̇w cosαw + ẋw= 0",
   PlotStyle → {Orange, Thick}, PlotRange → All,
```

```
      LabelStyle → Directive[fontsize, Black, Bold, ImageSize → imagesize],
      ImageSize → imagesize];
(*W2*)
PLOTW2 = Plot[Evaluate[r D[x[6][t]/.SOL, t]Sin[x[4][t]/.SOL]+
      D[x[2][t]/.SOL, t]], {t, t0, t1}, GridLines->Automatic,
      AxesLabel->{"t [s]", "w2"},
      PlotLabel->"quasi velocity w2 = r $\dot{\gamma}_w$ sin$\alpha_w$ + $\dot{y}_w$= 0 ",
      PlotStyle → {Orange, Thick}, PlotRange → All,
      LabelStyle → Directive[fontsize, Black, Bold, ImageSize → imagesize],
      ImageSize → imagesize];
(*W3*)
PLOTW3 = Plot[Evaluate[D[x[3][t]/.SOL, t]], {t, t0, t1},
      GridLines->Automatic, AxesLabel->{"t [s]", "w3"},
      PlotLabel- > "quasi velocity w3 = $\dot{z}_w$= 0 ",
      PlotStyle → {Orange, Thick}, PlotRange → All,
      LabelStyle → Directive[fontsize, Black, Bold, ImageSize → imagesize],
      ImageSize → imagesize];
(*W4*)
PLOTW4 = Plot[Evaluate[w[4][t]]/.SOL, {t, t0, t1},
      GridLines->Automatic, AxesLabel->{"t [s]", "w4[rad/s]"},
      PlotStyle → {Green, Thick}, PlotRange → All,
   PlotLabel->"quasi velocity w4 =$\dot{\beta}_w$ ", PlotRange → All,
      LabelStyle → Directive[fontsize, Black, Bold], ImageSize → imagesize] ;
(*W5*)
PLOTW5 = Plot[Evaluate[w[5][t]/.SOL], {t, t0, t1},
GridLines->Automatic, AxesLabel->{"t [s]", "w5 [rad/s]"},
      PlotStyle → {Green, Thick},
      PlotLabel->"quasi velocity w5 =$\dot{\alpha}_w$sin$\beta_w$ ",
      PlotRange → All, LabelStyle → Directive[fontsize, Black, Bold],
      ImageSize → imagesize] ;
(*W6*)
PLOTW6 = Plot[Evaluate[w[6][t]/.SOL], {t, t0, t1},
GridLines->Automatic, AxesLabel->{"t [s]", "w6 [rad/s]"},
      PlotStyle → {Green, Thick},
   PlotLabel->"quasi velocity w6 =$\dot{\gamma}_w$ + $\dot{\alpha}_w$cos$\beta_w$ ",
      PlotRange → All, LabelStyle → Directive[fontsize, Black, Bold],
      ImageSize → imagesize] ;
(*W7*)
PLOTW7 = Plot[Evaluate[w[7][t]]/.SOL, {t, t0, t1},
      GridLines->Automatic, AxesLabel->{"t [s]", "w6[rad/s]"},
      PlotStyle → {Green, Thick}, PlotRange → All,
   PlotLabel- > "quasi velocity w7 =$\dot{\alpha}_f$ + $\alpha_w$cos$\beta_f$ ",
      PlotRange → All,
      LabelStyle → Directive[fontsize, Black, Bold], ImageSize → imagesize] ;
```

```
(*W8*)
PLOTW8 = Plot[Evaluate[w[8][t]]/.SOL, {t, t0, t1},
PlotRange → All, PlotRange → All, GridLines->Automatic,
  AxesLabel->{"t [s]", "w8[rad/s]"}, PlotStyle → {Green, Thick},
   PlotRange → All,
PlotRange → All, PlotLabel->"quasi velocity w8", PlotRange → All,
  LabelStyle → Directive[fontsize, Black, Bold], ImageSize → imagesize];
(*W9*)
PLOTW9 = Plot[Evaluate[w[9][t]]/.SOL, {t, t0, t1},
PlotRange → All,
  PlotRange → All, PlotRange → All, PlotRange → All,
  GridLines->Automatic, AxesLabel->{"t [s]", "w9[rad/s]"},
  PlotStyle → {Green, Thick}, PlotRange → All,
PlotRange → All, PlotRange → All, PlotRange → All, PlotRange → All,
  PlotLabel->"quasi velocity w9", PlotRange → All,
  LabelStyle → Directive[fontsize, Black, Bold], ImageSize → imagesize];
(*W10*)
(*QUASI-VELOCITIES PLOTS*)


(*GENERAL COORDINATIONS PLOTS*)
(*X1*)
PLOTX1 = Plot[Evaluate[x[1][t]]/.SOL, {t, t0, t1},
  GridLines->Automatic, AxesLabel->{"t [s]", "xw [m]"} , PlotStyle →
   {Blue, Thick},
   PlotLabel- > "generalized coordinate x1 = xw", PlotRange → All,
  LabelStyle → Directive[fontsize, Black, Bold], ImageSize → imagesize];
(*X2*)
PLOTX2 = Plot[Evaluate[x[2][t]]/.SOL, {t, t0, t1},
  GridLines->Automatic, AxesLabel->{"t [s]", "yw [m]"} , PlotStyle →
   {Blue, Thick},
   PlotLabel- > "generalized coordinate x2 = yw",
   PlotRange → All,
  LabelStyle → Directive[fontsize, Black, Bold], ImageSize → imagesize];
(*X3*)
PLOTX3 = Plot[Evaluate[x[3][t]]/.SOL, {t, t0, t1},
  GridLines->Automatic, AxesLabel->{"t [s]", "zw [m]"} , PlotStyle →
   {Blue, Thick},
   PlotLabel- > "generalized coordinate x3 = zw", PlotRange → All,
  LabelStyle → Directive[fontsize, Black, Bold], ImageSize → imagesize];
```

```
(*X4*)
PLOTX4 = Plot[Evaluate[x[4][t]]/.SOL, {t, t0, t1},
  GridLines->Automatic, AxesLabel->{"t [s]", "αw [rad]"},
  PlotStyle → {Darker[Blue], Thick},
   PlotLabel- > "generalized coordinate x4 = αw(t)", PlotRange → All,
   LabelStyle → Directive[fontsize, Black, Bold], ImageSize → imagesize];
(*X5*)
PLOTX5 = Plot[Mod[(Evaluate[x[5][t]]/.SOL), 2π](*/Degree[°]*), {t, t0, t1},
  GridLines->Automatic, AxesLabel->{"t [s]", "βw [rad]"},
  PlotStyle → {Darker[Blue], Thick}, PlotRange → All,
   PlotLabel- > "generalized coordinate x6 = βw(t)", PlotRange → All,
   LabelStyle → Directive[fontsize, Black, Bold], ImageSize → imagesize];
(*X6*)
PLOTX6 = Plot[Mod[(Evaluate[x[6][t]]/.SOL), 2π](*/Degree[°]*), {t, t0, t1},
  GridLines->Automatic, AxesLabel->{"t [s]", "γw [rad]"},
  PlotStyle → {Darker[Blue], Thick},
   PlotLabel- > "generalized coordinate x6 = γw(t)", PlotRange → All,
   LabelStyle → Directive[fontsize, Black, Bold], ImageSize → imagesize];
(*x7*)
PLOTX7 = Plot[Mod[(Evaluate[x[7][t]]/.SOL), 2π](*/Degree[°]*), {t, t0, t1},
  GridLines->Automatic, AxesLabel->{"t [s]", "αf [rad]"},
  PlotStyle → {Darker[Blue], Thick},
   PlotLabel- > "generalized coordinate x7 = αf(t)", PlotRange → All,
   LabelStyle → Directive[fontsize, Black, Bold], ImageSize → imagesize];
(*X8*)
PLOTX8 = Plot[Mod[(Evaluate[x[8][t]]/.SOL), 2π], {t, t0, t1},
  GridLines->Automatic, AxesLabel->{"t [s]", "βb [rad]"},
  PlotStyle → {Darker[Blue], Thick}, PlotRange → All,
   PlotLabel- > "generalized coordinate x8 = βb(t)", PlotRange → All,
   LabelStyle → Directive[fontsize, Black, Bold], ImageSize → imagesize];
(*X9*)
PLOTX9 = Plot[Mod[(Evaluate[x[9][t]]/.SOL), 2π], {t, t0, t1},
  GridLines->Automatic, AxesLabel->{"t [s]", "γb [rad]"},
  PlotStyle → {Darker[Blue], Thick}, PlotRange → All,
   PlotLabel- > "generalized coordinate x9 = γb(t)", PlotRange → All,
   LabelStyle → Directive[fontsize, Black, Bold], ImageSize → imagesize];
(*X10*)
(*GENERAL COORDINATIONS PLOTS*)


(*TABLE OF TRUTH*)
TABLE = GraphicsGrid[{{PLOTW1, PLOTW2, PLOTW3},
   {PLOTW4, PLOTW5, PLOTW6}, {PLOTX4, PLOTX5, PLOTX6},
   {PLOTX7, PLOTW7, PLOTW8}, {PLOTX8, PLOTX9(*,PLOTX10*))}},
   Frame → All, ImageSize → imagesize * 1]
```

(*TABLE OF TRUTH*)

(*EXPORT TO FILE*)
(*Export ["TABLE.png", TABLE]; *)
(* EXPORT TO FILE *)

symbol	description	value	unit
g	gravity	9.81	m/s^2
r	wheel radius	0.3	m
h	hub length	0.12	m
c	crank length	0.137	m
p	pedal length	0.1	m
f	frame length	0.65	m
mw	wheel mass	2.4	kg
mh	hub mass	0.1	kg
mc	crank mass	0.3	kg
mp	pedal mass	0.1	kg
mf	frame mass	0.7	kg
mb	frame mass	0.004	kg
x0[1]	x_w	0.	m
x0[2]	y_w	0.	m
x0[3]	z_w	0.	m
x0[4]	initial position α_w	0.	$\backslash^o$
x0[5]	β_w	90	$\backslash^o$
x0[6]	γ_w	0.	$\backslash^o$
x0[7]	α_f	0.	$\backslash^o$
x0[8]	α_b	0.	$\backslash^o$
x0[9]	β_b	0.	$\backslash^o$
dx0[4]	initial velocity $\dot{\alpha}_w$	0.	rad
dx0[5]	$\dot{\beta}_w$	0.	rad
dx0[6]	$\dot{\gamma}_w$	3.	rad
dx0[7]	$\dot{\alpha}_f$	-2.76923	rad
dx0[7]	$\dot{\alpha}_b$	0	rad
dx0[5]	$\dot{\beta}_b$	-1.05882	rad
—	linear velocity	3.24	rad
dt	max step size	$\frac{1}{10000}$	s
t0	start time	0	s
t1	stop time	2.	s
Δt	motion time	2.	s

YYYY
DONE

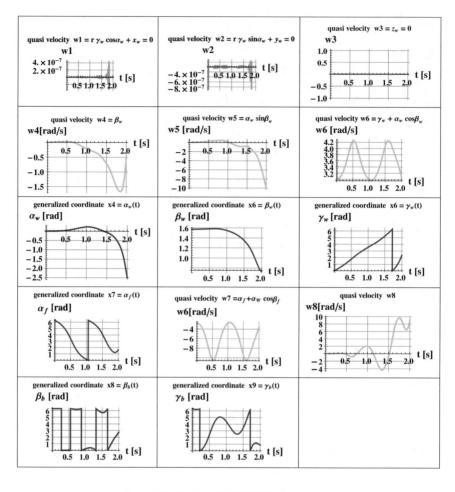

$$x10jg = \left(2\pi + \text{ArcSec}\left[\frac{2ti\sqrt{c^2+f^2-2cf\text{Cos}[x[6][t]-x[7][t]]}}{c^2+f^2-th^2+ti^2-2cf\text{Cos}[x[6][t]-x[7][t]]}\right]\right.$$
$$\left.- \text{ArcTan}[f - c\text{Cos}[x[6][t] - x[7][t]], c\text{Sin}[x[6][t] - x[7][t]] - x[6][t] + x[7][t])\right/.$$

SOL;

PLOTX10jg = Plot[Mod[x10jg, 2π], {t, t0, t1},

GridLines->Automatic, AxesLabel->{"t [s]", "x10ψ_{tir}[rad]"},

   PlotStyle $\to$ {Darker[Blue], Thick}, PlotRange $\to$ All,

PlotLabel- > "generalized coordinate x10 = $\psi_{\text{tir}}(t)$", PlotRange $\to$ All,

   LabelStyle $\to$ Directive[fontsize, Black, Bold], ImageSize $\to$ imagesize]

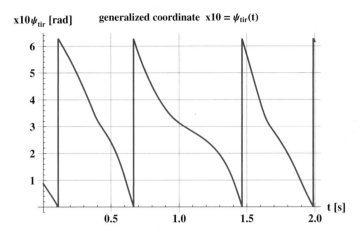

$$x[10][t] = \left(2\pi + \text{ArcSec}\left[\frac{2\text{ti}\sqrt{c^2+f^2-2cf\text{Cos}[x[6][t]-x[7][t]]}}{c^2+f^2-\text{th}^2+\text{ti}^2-2cf\text{Cos}[x[6][t]-x[7][t]]}\right]\right.$$
$$\left. - \text{ArcTan}[f - c\text{Cos}[x[6][t] - x[7][t]], c\text{Sin}[x[6][t] - x[7][t]]] - x[6][t] + x[7][t]\right)$$

```
(*3 − D TRAJECTORY*)
(*3 − D TYRE*)
TIREPP3D = ParametricPlot3D [Flatten[rOT]/.SOL, {t, t0, t1},
    AspectRatio → Automatic, AxesLabel → {"x_T", "y_T", "z_T"},
PlotStyle → {Yellow, Thick}, PlotRange → All,
    LabelStyle → Directive[fontsize, Black, Bold], ImageSize → imagesize];
(*3 − D TYRE*)


(*3 − D WHEEL*)
HUBPP3D = ParametricPlot3D [Flatten[rOH]/.SOL, {t, t0, t1},
    AspectRatio → Automatic, AxesLabel → {"x_H", "y_H", "z_H"},
PlotStyle → {Magenta, Dashed, Thick, Thick}, PlotRange → All,
    LabelStyle → Directive[fontsize, Black, Bold], ImageSize → imagesize];
PRDALRPP3D = ParametricPlot3D [Flatten[rOPR]/.SOL, {t, t0, t1},
    AspectRatio → Automatic, AxesLabel → {"x_PR", "y_PR", "z_PR"},
PlotStyle → {Blue, Dashed}, PlotRange → All,
    LabelStyle → Directive[fontsize, Black, Bold], ImageSize → imagesize];
PRDALLPP3D = ParametricPlot3D [Flatten[rOPL]/.SOL, {t, t0, t1},
    AspectRatio → Automatic, AxesLabel → {"x_PL", "y_PL", "z_PL"},
PlotStyle → {Red, Dashed}, PlotRange → All,
    LabelStyle → Directive[fontsize, Black, Bold], ImageSize → imagesize];
RIM3D =
{Black, CapForm[None],
Rotate[Rotate[Tube[{{rOH[[1, 1]], rOH[[2, 1]], rOH[[3, 1]] − 0.01},
```

```
    {rOH[[1, 1]], rOH[[2, 1]], rOH[[3, 1]] + 0.01}}, r], x[4][t],
{0, 0, 1}, Flatten[rOH]], x[5][t], Flatten[Rα[4].Rβ[4].{1, 0, 0}], Flatten[rOH]]};
(* 3-D CRANKSET *)
CRANKSET3D = {Thick, Blue, Line[{Flatten[rOPR], Flatten[rOCR]}],
    Black, Line[{Flatten[rOCR], Flatten[rOHR2], Flatten[rOHL2], Flatten[rOCL]}],
Red, Line[{Flatten[rOCL], Flatten[rOPL]}]};
WHEEL3D[t_]:=Graphics3D[{CRANKSET3D, RIM3D}];
(*3 − D WHEEL*****)
```

```
(*3 − D FRAME*)
FRAMEPP3D = ParametricPlot3D [Flatten[rOS]/.SOL, {t, t0, t1},
    AspectRatio →  Automatic, AxesLabel →  {"x_f", "y_f", "z_f"},
PlotStyle → {Green, Dashed, Thick, Thick}, PlotRange → All,
    LabelStyle → Directive[fontsize, Black, Bold], ImageSize → imagesize];
FRAME3D[t_]:=Graphics3D[{Thick, Line[{Flatten[rOH], Flatten[rOS]}]}];
(*3 − D FRAME*****)
```

```
(*3 − D BODY*)
BODYPP3D = ParametricPlot3D [Flatten[rOB]/.SOL, {t, t0, t1},
    AspectRatio → Automatic, AxesLabel → {"x_b", "y_b", "z_b"},
PlotStyle → {Orange, Dashed, Thick, Thick}, PlotRange → All,
    LabelStyle → Directive[fontsize, Black, Bold], ImageSize → imagesize];
BODY3D[t_]:=Graphics3D[{Thick, Line[{Flatten[rOSR], Flatten[rOSL]}],
    Line[{Flatten[rOS], Flatten[rOB]}]}];
KOLO =
{Blue, CapForm[None],
Rotate[Rotate[Tube[{{rOSR[[1, 1]], rOSR[[2, 1]], rOSR[[3, 1]] − 0.003},
    {rOSR[[1, 1]], rOSR[[2, 1]], rOSR[[3, 1]] + 0.003}}, th],
x[4][t], {0, 0, 1}, Flatten[rOSR]], x[5][t], Flatten[Rα[4].Rβ[4].{1, 0, 0}],
    Flatten[rOSR]]};
TEST[t_]:=Graphics3D[{KOLO}];
(*3 − D BODY*)
```

```
(*3 − D TIBIA*)
KNEERPP3D = ParametricPlot3D [Flatten[rOKR]/.SOL, {t, t0, t1},
    AspectRatio → Automatic, AxesLabel → {"x_b", "y_b", "z_b"},
PlotStyle → {Blue, Dashed, Thick, Thick}, PlotRange → All,
  "LabelStyle → Directive[fontsize, Black, Bold], ImageSize → imagesize];
TIBIA3D[t_]:=Graphics3D[{Thick, Blue, Line[{Flatten[rOPR2],
    Flatten[rOKR]}]}];
(*3 − D TIBIA*)
```

```
(*AREA FITTING TRAJECTORY*)
xMAX = Max[Table[x[1][t]/. SOL, {t, t0, t1, t1/1000}]];
yMAX = Max[Table[x[2][t]/. SOL, {t, t0, t1, t1/1000}]];
xMIN = Min[Table[x[1][t]/. SOL, {t, t0, t1, t1/1000}]];
yMIN = Min[Table[x[2][t]/. SOL, {t, t0, t1, t1/1000}]];
RANGE = {{xMIN − (f + b), xMAX + (f + b)},
    {yMIN − (f + b), yMAX + (f + b)}, {1.1(f + r + b), −1.1(f − r + b)}};
AREA =
Graphics3D[
{Lighter[Lighter[Green]],

Polygon[{{RANGE[[1, 2]], RANGE[[2, 2]], 0}, {RANGE[[1, 1]], RANGE[[2, 2]], 0},
    {RANGE[[1, 1]], RANGE[[2, 1]], 0},
{RANGE[[1, 2]], RANGE[[2, 1]], 0}}]}];
(*AREA FITTING TRAJECTORY*)


(*3 − D TRAJECTORY + WHEEL*)
TRAJECTORY = Show[HUBPP3D, TIREPP3D, PRDALRPP3D,
    PRDALLPP3D, FRAMEPP3D, BODYPP3D, KNEERPP3D, AERA/.SOL/.t
        → t1,
WHEEL3D[t]/.SOL/.t → 0, WHEEL3D[t]/.SOL/.t → t1/4,
    WHEEL3D[t]/.SOL/.t → t1/3, WHEEL3D[t]/.SOL/.t → t1/2,
    WHEEL3D[t]/.SOL/.t → t1/1,
FRAME3D[t]/.SOL/.t → 0, FRAME3D[t]/.SOL/.t → t1/4,
    FRAME3D[t]/.SOL/.t → t1/3, FRAME3D[t]/.SOL/.t → t1/2,
    FRAME3D[t]/.SOL/.t → t1/1,
BODY3D[t]/.SOL/.t → 0, BODY3D[t]/.SOL/.t → t1/4,
    BODY3D[t]/.SOL/.t → t1/3, BODY3D[t]/.SOL/.t → t1/2,
    BODY3D[t]/.SOL/.t → t1/1,
TIBIA3D[t]/.SOL/.t → 0, TIBIA3D[t]/.SOL/.t → t1/4,
    TIBIA3D[t]/.SOL/.t → t1/3, TIBIA3D[t]/.SOL/.t → t1/2,
    TIBIA3D[t]/.SOL/.t → t1/1,
TEST[t]/.SOL/.t → 0, TEST[t]/.SOL/.t → t1/4,
    TEST[t]/.SOL/.t → t1/3, TEST[t]/.SOL/.t → t1/2,
    TEST[t]/.SOL/.t → t1/1,
PlotRange → RANGE, AspectRatio → Automatic, Boxed → False,
    AxesLabel → {"x [m]", "y [m]", "z [m]"},
    LabelStyle → Directive[fontsize, Black, Bold],
ImageSize → imagesize ∗ 2.4, ViewPoint → {0, 8, 1}]

TRAJECTORY = Show[HUBPP3D, TIREPP3D, PRDALRPP3D, PRDALLPP3D,
    FRAMEPP3D, BODYPP3D, KNEERPP3D, AERA/.SOL/.t → t1,
WHEEL3D[t]/.SOL/.t → 0, WHEEL3D[t]/.SOL/.t → t1/40,
    WHEEL3D[t]/.SOL/.t → t1/30, WHEEL3D[t]/.SOL/.t → t1/20,
```

WHEEL3D[t]/.SOL/.t → t1/10,
FRAME3D[t]/.SOL/.t → 0, FRAME3D[t]/.SOL/.t → t1/40,
   FRAME3D[t]/.SOL/.t → t1/30, FRAME3D[t]/.SOL/.t → t1/20,
   FRAME3D[t]/.SOL/.t → t1/10,
BODY3D[t]/.SOL/.t → 0, BODY3D[t]/.SOL/.t → t1/40,
   BODY3D[t]/.SOL/.t → t1/30, BODY3D[t]/.SOL/.t → t1/20,
   BODY3D[t]/.SOL/.t → t1/10,
TIBIA3D[t]/.SOL/.t → 0, TIBIA3D[t]/.SOL/.t → t1/40,
   TIBIA3D[t]/.SOL/.t → t1/30, TIBIA3D[t]/.SOL/.t → t1/20,
   TIBIA3D[t]/.SOL/.t → t1/10,
TEST[t]/.SOL/.t → 0, TEST[t]/.SOL/.t → t1/40,
   TEST[t]/.SOL/.t → t1/30, TEST[t]/.SOL/.t → t1/20,
   TEST[t]/.SOL/.t → t1/10,
PlotRange → RANGE, AspectRatio → Automatic, Boxed → False,
   AxesLabel → {"x [m]", "y [m]", "z [m]"},
   LabelStyle → Directive[fontsize, Black, Bold],
ImageSize → imagesize ∗ 2.4, ViewPoint → {0, 8, 1}]
(*3 − D TRAJECTORY + WHEEL*)

(*EXPORT TO FILE*)
(*Export["TRAJECTORY.png", TRAJECTORY]; *)
(*EXPORT TO FILE*********************)
(*3 − D TRAJECTORY*)

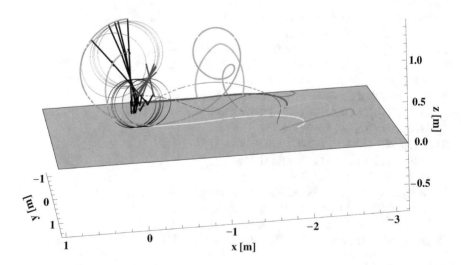

References

1. Nikravesh, P. (1998). *Computer-aided analysis of mechanical systems*. Engelwood Cliffs: Prentice-Hall.
2. Diebel, J. (2006). *Representing attitude: Euler angles, unit quaternions, and rotation vectors*. Stanford University Press.
3. Weisstein, E. (2015). Sensors *15*(3), http://mathworld.wolfram.com/EulerAngles.html.
4. Janota, A., Simak, V., Nemec, D., & Hrbĉek, J. (2015). Improving the precision and speed of Euler angles computation from low-cost rotation sensor data. *Sensors (Basel)*, *15*, 7016–7039.

Printed in the United States
By Bookmasters